China College Entrance Examination Mathematics Questions

How Chinese Students Achieve High Score in Mathematics Series

Collected & Translated by Haiqing Hua

haiqinghua@yahoo.com

Haiqing Hua | LinkedIn

Table of Contents

2022年普通高等学校招生全国统一考试（新高考I卷）

数　　学

2022 National Unified Entrance Examination for Ordinary Higher Education (New Gaokao I) Mathematics Note: *Gaokao* is the Chinese national college entrance examination.

注意事项：

1. 答卷前，考生务必将自己的姓名、考生号等填写在答题卡和试卷指定位置上。

2. 回答选择题时，选出每小题答案后，用铅笔把答题卡上对应题目的答案标号涂黑。如需改动，用橡皮擦干净后，再选涂其他答案标号。回答非选择题时，将答案写在答题卡上。写在本试卷上无效。

3. 请保持答题卡的整洁。考试结束后，将试卷和答题卡一并交回。

Notes:

1. Before answering, candidates must fill in their name, candidate number, and other information in the designated areas on the answer sheet and the exam paper.
2. When answering multiple-choice questions, fill in the corresponding answer on the answer sheet with a pencil after selecting the answer for each question. If you need to change your answer, erase the original answer completely before filling in a new one. For non-multiple-choice questions, write the answers on the answer sheet. Answers written on the exam paper will not be valid.
3. Please keep the answer sheet clean. After the exam, turn in both the exam paper and the answer sheet.

一、选择题：本题共8小题，每小题5分，共40分。在每小题给出的四个选项中，只有一项是符合题目要求的。

Section I: Multiple Choice Questions

This section contains 8 questions, each worth 5 points, for a total of 40 points. For each question, choose the one best answer from the four options provided.

1. If set $M=\{x\mid\sqrt{x}<4\}$，$N=\{x\mid 3x\geqslant 1\}$，则 $M\cap N=$

A. $\{x\mid 0\leqslant x<2\}$　　B. $\{x\mid \frac{1}{3}\leqslant x<2\}$

C. $\{x\mid 3\leqslant x<16\}$　　D. $\{x\mid \frac{1}{3}\leqslant x<16\}$

ANSWER: D

ANALYSIS: set $M=\{x\mid 0\leqslant x<16\}$,

set $N=\{x\mid x\geqslant\frac{1}{3}\}$，$M\cap N=\{x\mid \frac{1}{3}\leqslant x<16\}$. so we select D

2. If $i(1-z)=1$, then $z+\bar{z}=$

A. -2　　B. -1　　C. 1　　D. 2

Answer: D

Analysis:

【解析】对原式两边同时乘以i得：$z-1=i$，即$z=1+i$，所以$\bar{z}=1-i$，即$z+\bar{z}=2$.

Analysis: Multiplying both sides of the original equation by , we get: $z-1=i$, which means $z = 1 + i$. Therefore, three = 1 - 1, which means: + 3 = 2. So we select D

3. If in $\triangle ABC$ the point D on the side AB $BD = 2DA$.

Let $\overrightarrow{CA} = m$, $\overrightarrow{CD} = n$, then $\overrightarrow{CB} =$

A. $3m - 2n$ B. $-2m + 3n$ C. $3m + 2n$ D. $2m + 3n$

Answer: D

Analysis: Because $\overrightarrow{CB} = \overrightarrow{CA} \quad \overrightarrow{AB} \quad \overrightarrow{CA}$ and because

$\overrightarrow{AD} = \overrightarrow{CD} \quad ($, so $\overrightarrow{CB} = -2\overrightarrow{CA} + 3\overrightarrow{CD}$, which means

$\overrightarrow{CB} = -2m + 3n$. Therefore, we choose B

4. 南水北调工程缓解了北方一些地区水资源短缺问题，其中一部分水蓄入某水库，已知该水库水位为海拔148.5m时，相应水面的面积为140km^2；水位为海拔157.5m时，相应水面的面积为180km^2，将该水库在这两个水位间的形状看作一个棱台，则该水库水位从海拔148.5m上升到157.5m时，增加的水量约为($\sqrt{7} \approx 2.65$)

English Translation:

The South-to-North Water Diversion Project has alleviated water shortages in some northern regions. Part of this water is stored in a certain reservoir. It is known that when the water level of this reservoir is 148.5 meters above sea level, the corresponding water surface area is 140 square kilometers; when the water level is 157.5 meters above sea level, the corresponding water surface area is 180 square kilometers. If the shape of the reservoir between these two water levels is considered a frustum, then the amount of water increased when the water level of the reservoir rises from 148.5 meters to 157.5 meters is approximately($\sqrt{7} \approx 2.65$) **[unit of volume].**

A. $1.0 \times 10^9 m^3$ B. $1.2 \times 10^9 m^3$

C. $1.4 \times 10^9 m^3$ D. $1.6 \times 10^9 m^3$

Answer: C

【解析】由题意 $S_1 = 140km^2$，$S_2 = 180km^2$，$h = (157.5 - 148.5)km = 9km$，代入棱台体积 $V = \frac{1}{3}(S_1 + S_2 + \sqrt{S_1 S_2})h$，公式可得：$V \approx 1.4 \times 10^9 m^3$．故选C．

Analysis: Given S_1 = 140 km^2, S_2 = 180 km^2, and h = (157.5 - 148.5) km = 9 km, substitute these values into the formula for the volume of a frustum:

$V = \frac{1}{3}(S_1 + S_2 + \sqrt{S_1 S_2})h$, We get: $V \approx 1.4 \times 10^9 m^3$. Therefore, the answer is C.

5. 从2至8的7个整数中随机取2个不同的数，则这2个数互质的概率为

5. From the 7 integers from 2 to 8, randomly select 2 different numbers. What is the probability that these 2 numbers are coprime?

A. $\frac{1}{6}$ B. $\frac{1}{3}$ C. $\frac{1}{2}$ D. $\frac{2}{3}$

Answer: D

Analysis:

【解析】总事件数共 $C_7^2=\frac{7\times 6}{2}=21$，

第一个数取2时，第二个数可以是3,5,7；

第一个数取3时，第二个数可以是4,5,7,8；

第一个数取4时，第二个数可以是5,7；

第一个数取5时，第二个数可以是6,7,8；

第一个数取6时，第二个数可以是7；

第一个数取7时，第二个数可以是8；

所以 $P=\frac{3+4+2+3+1+1}{21}=\frac{14}{21}=\frac{2}{3}$．

Analysis: The total number of possible outcomes is C(7, 2) = 21.
When the first number is 2, the second number can be 3, 5, or 7. When the first number is 3, the second number can be 4, 5, 7, or 8. When the first number is 4, the second number can be 5 or 7. When the first number is 5, the second number can be 6, 7, or 8. When the first number is 6, the second number can be 7. When the first number is 7, the second number can be 8.
Therefore, the probability P = (3 + 4 + 2 + 3 + 1 + 1) / 21 = 14 / 21 = 2/3.

6．记函数 $f(x)=\sin(\omega x+\frac{\pi}{4})+b(\omega>0)$ 的最小正周期为 T．若 $\frac{2}{3}\pi<T<\pi$，且 $y=f(x)$ 的函数图象关于点 $(\frac{3\pi}{2},2)$ 中心对称，则 $f(\frac{\pi}{2})=$

6. Given function $f(x)=\sin(\omega x+\frac{\pi}{4})+b(\omega>0)$ with the smallest positive period T. If $\frac{2}{3}\pi<T<\pi$, & the graph of y = f(x) is symmetric about the point $(\frac{3\pi}{2},2)$, then f(π) =

A．1　　B．$\frac{3}{2}$　　C．$\frac{5}{2}$　　D．3

Answer: A

【解析】$\omega=\frac{2\pi}{T}\in(2,3)$，$y=f(x)$ 的函数图象关于点 $(\frac{3\pi}{2},2)$ 中心对称，则有 $b=2$，且

Analysis: Since $\omega=\frac{2\pi}{T}\in(2,3)$, the graph of y = f(x) is symmetric about the point $(\frac{3\pi}{2},2)$, so b = 2. And $f(\frac{3\pi}{2})=2$, so $\sin(\frac{3\pi}{2}\omega+\frac{\pi}{4})+2=2$, then $\frac{3\pi}{2}\omega+\frac{\pi}{4}=2k\pi, k\in\mathbf{Z}$; we get $\omega=\frac{8k-1}{6}$, by $\omega\in(2,3)$ we get

$k=2$, $\omega=\frac{5}{2}$, so $f(\frac{\pi}{2})=\sin(\frac{5}{2}\cdot\frac{\pi}{2}+\frac{\pi}{4})+2=-1+2=1$.

7, Let $a=0.1e^{0.1}$, $b=\frac{1}{9}$, $c=-\ln 0.9$, **then**

A. $a<b<c$ B. $c<b<a$ C. $c<a<b$ D. $a<c<b$

Answer: C

Analysis: let $a=xe^{x}$, $b=\frac{x}{1-x}$, $c=-\ln(1-x)$.

① $\ln a-\ln b=x+\ln x-[\ln x-\ln(1-x)]$.

$y=x+\ln(1-x), x\in(0,0.1]$; $y'=1-\frac{1}{1-x}=\frac{-x}{1-x}<0$,

Therefore, 7=<0, so $\ln a-\ln b\leq 0$, so b>a

② $a-c=xe^{x}+\ln(1-x), x\in(0,0.1]$,

$y'=xe^{x}+e^{x}-\frac{1}{1-x}=\frac{(1+x)(1-x)e^{x}-1}{1-x}$,

Let so $k(x)=(1+x)(1-x)e^{x}-1$, so $k'(x)=(1-x^2-2x)e^{x}>0$, so $k(x)>k(0)>0$, so $y'>0$, so a-c>0 so a>c.

8．已知正四棱锥的侧棱长为l，其各顶点都在同一球面上．若该球的体积为36π，且$3\leq l\leq 3\sqrt{3}$，则该正四棱锥体积的取值范围是

8. Given a regular square pyramid with a slant height of l, whose vertices all lie on the same sphere. If the volume of the sphere is 36π, and $3\leq l\leq 3\sqrt{3}$ **, then the range of the volume of the pyramid is**

A. $[18,\frac{81}{4}]$ B. $[\frac{27}{4},\frac{81}{4}]$ C. $[\frac{27}{4},\frac{64}{3}]$ D. $[18,27]$

Answer: C

【解析】记三棱锥高与侧棱夹角为θ，高为h，底面中心到各顶点的距离为m，

Analysis: Let θ be the angle between the height and slant height of the pyramid, h be the height, m be the distance from the center of the base to each vertex, and a be the side length of the base.

$\cos\theta=\frac{3^2+l^2-3^2}{2\times 3\times l}=\frac{l}{6}\in[\frac{1}{2},\frac{\sqrt{3}}{2}]$, then

$l=6\cos\theta$, $m=l\cdot\sin\theta=6\sin\theta\cos\theta$,

$h=\frac{m}{\tan\theta}=\frac{6\sin\theta\cos\theta}{\frac{\sin\theta}{\cos\theta}}=6\cos^2\theta$, $S_{底}=\frac{1}{2}\times 2m\times 2m=2m^2$,

so

$V=\frac{1}{3}S_{底}\cdot h=\frac{1}{3}\times 2m^2h=144(\sin\theta\cos^2\theta)^2$，

Let

$y=\sin\theta\cos^2\theta=\sin\theta(1-\sin^2\theta)=x(1-x^2)=-x^3+x, x=\sin\theta\in[\frac{1}{2},\frac{\sqrt{3}}{2}]$

$y'=-3x^2+1$，故 $x\in[\frac{1}{2},\frac{\sqrt{3}}{3})$，$y'<0$，$x\in(\frac{\sqrt{3}}{3},\frac{\sqrt{3}}{2}]$，$y'>0$，

Which means $V_{max}=144y_{max}^2=144\times[\frac{\sqrt{3}}{3}\times(\frac{\sqrt{6}}{3})^2]^2=\frac{64}{3}$，

$V_{min}=144\times(\frac{\sqrt{3}}{2}\times(\frac{1}{2})^2)^2=\frac{27}{4}$.

二、选择题：本题共 4 小题，每小题 5 分，共 20 分。在每小题给出的选项中，有多项符合题目要求。全部选对的得 5 分，部分选对的得 2 分，有选错的得 0 分。

Section II: Multiple Choice Questions

This section contains 4 questions, each worth 5 points, for a total of 20 points. For each question, there may be multiple correct answers. Selecting all correct answers earns 5 points, partially correct answers earn 2 points, and any incorrect selections earn 0 points.

9．已知正方体 $ABCD-A_1B_1C_1D_1$，则

A．直线 BC_1 与 DA_1 所成的角为 90°

B．直线 BC_1 与 CA_1 所成的角为 90°

C．直线 BC_1 与平面 BB_1D_1D 所成的角为 45°

D．直线 BC_1 与平面 $ABCD$ 所成的角为 45°

9. Given a cube ABCD-A'B'C'D', then

- A. The angle between lines BC and DA' is 90°.
- B. The angle between lines BC and CA' is 90°.
- C. The angle between line BC and plane BB'D'D is 45°.
- D. The angle between line BC and plane ABCD is 45°.

Answer: A B C

【解析】在正方体 $ABCD-A_1B_1C_1D_1$ 中，因为 $BC_1\perp B_1C$，$BC_1\perp A_1B_1$，所以 $BC_1\perp$ 平面 A_1B_1CD，所以 $BC_1\perp DA_1$，$BC_1\perp CA_1$，故选项 A，B 均正确；

设 $A_1C_1\cap B_1D_1=O$，因为 $A_1C_1\perp$ 平面 BB_1D_1D，所以直线 BC_1 与平面 BB_1D_1D 所成的角为 $\angle C_1BO$，在直角 $\triangle C_1BO$ 中，$\sin\angle C_1BO=\frac{C_1O}{BC_1}=\frac{1}{2}$，故 $\angle C_1BO=30°$，故选项 C 错误；

直线 BC_1 与平面 $ABCD$ 所成的角为 $\angle C_1BC=45°$，故选项 D 正确．综上，答案选 ABD.

Analysis: In the cube $ABCD-A_1B_1C_1D_1$, since $BC_1\perp B_1C$，$BC_1\perp A_1B_1$，so that $BC_1\perp$ plane A_1B_1CD，so $BC_1\perp DA_1$，$BC_1\perp CA_1$，therefore A, B are all correct:

Let $A_1C_1 \cap B_1D_1 = O$, since the angle between $A_1C_1 \perp$ Plane BB_1D_1D is $\angle C_1BO$. In the right triangle $\triangle C_1BO$ $\sin\angle C_1BO = \frac{C_1O}{BC_1} = \frac{1}{2}$, so $\angle C_1BO = 30°$, so C is not correct. Since the angle between line BC_1 and plane $ABCD$ is $\angle C_1BC = 45°$ so D is correct.
Therefore, the answer is ABD

10．已知函数 $f(x)=x^3-x+1$，则

A．$f(x)$ 有两个极值点

B．$f(x)$ 有三个零点

C．点 $(0,1)$ 是曲线 $y=f(x)$ 的对称中心

D．直线 $y=2x$ 是曲线 $y=f(x)$ 的切线

10. Given the function $f(x)=x^3-x+1$**, then**

- A. f(x) has two extreme points.
- B. f(x) has three zeros.
- C. The point (0, 1) is the center of symmetry of the curve y = f(x).
- D. The line y = 2x is a tangent line to the curve y = f(x).

Answer: AC

【解析】$f'(x)=3x^2-1$，所以 $f(x)$ 有两个极值点 $-\frac{\sqrt{3}}{3}$ 与 $\frac{\sqrt{3}}{3}$，又 $f(\frac{\sqrt{3}}{3})=1-\frac{2\sqrt{3}}{9}>0$，

所以 $f(x)$ 只有一个零点；由 $f(x)+f(-x)=2$ 可知，点 $(0,1)$ 是曲线 $y=f(x)$ 的对称中心；曲线 $y=f(x)$ 在点 $(1,1)$ 处的切线方程为 $y=2x-1$，所以答案选 AC.

Analysis: f'(x) = $3x^2$ - 1, so f(x) has two extreme points

$-\frac{\sqrt{3}}{3}$ 与 $\frac{\sqrt{3}}{3}$，又 $f(\frac{\sqrt{3}}{3})=1-\frac{2\sqrt{3}}{9}>0$

Also
so f(x) has only one zero. Since f(x) + f(-x) = 2, the point (0, 1) is the center of symmetry of the curve y = f(x). The equation of the tangent line to the curve y = f(x) at point (1, 1) is y = 2x - 1, so the answer is AC.

已知 O 为坐标原点，点 $A(1,1)$ 在抛物线 C：$x^2=2py$（$p>0$）上，过点 $B(0,-1)$ 的直线交 C 于 P,Q 两点，则

Given: O is the origin, point A(1,1) is on the parabola C: x^2 = 2py (p > 0). A straight line passing through point B(0, -1) intersects C at points P and Q. Then...

A．C 的准线为 $y=-1$　B．直线 AB 与 C 相切

C．$|OP|\cdot|OQ|>|OA|^2$　D．$|BP|\cdot|BQ|>|BA|^2$

Answer: BCD

【解析】由题意可知：$1=2p$，所以抛物线 C：$x^2=y$，故 C 的准线为 $y=-\frac{1}{4}$，故

A is not correct.

Analysis: From the given information, we know: 1 = 2p, so the parabola C：$x^2=y$，Therefore, the directrix of C is y = -1/4. So A is not Correct.

由 $y' = 2x$ 得曲线 C 在点 $A(1, 1)$ 处的切线斜率为 2，所以切线方程为 $y = 2x - 1$，故直线 AB 与 C 相切；

From the equation y' = 2x, we get that the slope of the tangent line to curve C at point A(1, 1) is 2. Therefore, the equation of the tangent line is y = 2x - 1. Hence, the line AB is tangent to C.

过点 $B(0, -1)$ 的直线设为 $y = kx - 1$，交 C 于 P, Q 两点的坐标分别设为 $P(x_1, y_1), Q(x_2, y_2)$，

Let the equation of the line passing through point B(0, -1) be y = kx - 1. Let the coordinates of the intersection points of this line with parabola C be $P(x_1, y_1)$ and $Q(x_2, y_2)$.

联立直线与 C 方程可得 $\begin{cases} x^2 = y \\ y = kx - 1 \end{cases} \Rightarrow x^2 - kx + 1 = 0$，所以有 $x_1 + x_2 = k$，$x_1 x_2 = 1$，

Combining the equations of the line and parabola C, we get:

$\begin{cases} x^2 = y \\ y = kx - 1 \end{cases} \Rightarrow x^2 - kx + 1 = 0$, so $x_1 + x_2 = k$, $x_1 x_2 = 1$, and $|OA|^2 = 2$,

so C is correct.

$|BP| \cdot |BQ| = \overrightarrow{BP} \cdot \overrightarrow{BQ} = (x_1, y_1 + 1) \cdot (x_2, y_2 + 1) = x_1 x_2 + y_1 y_2 + y_1 + y_2 + 1 = k^2 + 1 > 5$,

And $|BA|^2 = 5$, so the answer is B C D

12. 已知函数 $f(x)$ 及其导函数 $f'(x)$ 的定义域均为 **R**，记 $g(x) = f'(x)$ ．若 $f(\frac{3}{2} - 2x)$，$g(2 + x)$ 均为偶函数，则

Given: Functions f(x) and its derivative $f'(x)$ both have a domain of R. Let g(x) = f'(x). If f(3/2 - 2x) and g(2 + x) are both even functions, then...

A. $f(0) = 0$　　B. $g(-\frac{1}{2}) = 0$　　C. $f(-1) = f(4)$　　D. $g(-1) = g(2)$

Answer: B C

【解析】由 $f(\frac{3}{2} - 2x)$ 为偶函数可知 $f(x)$ 关于直线 $x = \frac{3}{2}$ 对称，

由 $g(2 + x)$ 为偶函数可知：$g(x)$ 关于直线 $x = 2$ 对称，

Analysis:

- Since f(3/2 - 2x) is an even function, f(x) is symmetric about the line x = 3/4.
- Since g(2 + x) is an even function, g(x) is symmetric about the line x = -2.

结合 $g(x) = f'(x)$，根据 $g(x)$ 关于直线 $x = 2$ 对称可知 $f(x)$ 关于点 $(2, t)$ 对称，

Combining g(x) = f'(x) with the fact that g(x) is symmetric about the line x = 2, we can conclude that f(x) is symmetric about the point (2, 1).

根据 $f(x)$ 关于直线 $x = \frac{3}{2}$ 对称可知：$g(x)$ 关于点 $(\frac{3}{2}, 0)$ 对称，

Based on the fact that f(x) is symmetric about the line x = 3/2, we can conclude that g(x) is symmetric about the point (3/2, 0).

综上，函数 $f(x)$ 与 $g(x)$ 均是周期为 2 的周期函数，所以有 $f(0) = f(2) = t$，所以 A 不正确；

In summary, both functions f(x) and g(x) have a period of 2. Therefore, f(0) = f(2) = t, so A is incorrect.

$f(-1) = f(1)$，$f(4) = f(2)$，$f(1) = f(2)$，故 $f(-1) = f(4)$，

So, C is correct.

$g(-\frac{1}{2})=g(\frac{3}{2})=0$， $g(-1)=g(1)$，

Therefore, B is correct.

And $g(1)+g(2)=0$，所以 $g(-1)+g(2)=0$， so D is not correct.

So, we select B C

三、填空题：本题共 4 小题，每小题 5 分，共 20 分。

Section III: Fill in the blanks. There are 4 questions in this section, each worth 5 points, for a total of 20 points.

13．$(1-\frac{y}{x})(x+y)^8$的展开式中x^2y^6的系数为________（用数字作答）.

【答案】-28

Question 13: Find the coefficient of x x^2y^6 in the expansion of $(1-\frac{y}{x})(x+y)^8$

Answer: -28

【解析】原式等于$(x+y)^8-\frac{y}{x}(x+y)^8$，由二项式定理，其展开式中$x^2y^6$的系数为

$C_8^2-C_8^3=-28$.

Analysis:

The original expression is equal to $(x+y)^8-\frac{y}{x}(x+y)^8$. Using the binomial theorem, the coefficient of x^2y^6 in the expansion is $C_8^2-C_8^3=-28$.

14．写出与圆$x^2+y^2=1$和$(x-3)^2+(y-4)^2=16$都相切的一条直线的方程________.

Question 14:

Write the equation of a straight line that is tangent to both the circle $x^2 + y^2 = 1$ and the circle $(x-3)^2 + (y-4)^2 = 16$.

【答案】$x=-1$，或$y=\frac{7}{24}x-\frac{25}{24}$，或$y=-\frac{3}{4}x+\frac{5}{4}$（答对其中之一即可）

Answer: $x=-1$ **or** $y=\frac{7}{24}x-\frac{25}{24}$ **or** $y=-\frac{3}{4}x+\frac{5}{4}$

(Any one of these answers is correct)

【解析】由图可得，两圆外切，且均与直线l_1：$x=-1$相切，另过两圆圆心的直线l的方程为$y=\frac{4}{3}x$，可得l与l_1交点为$P(-1,-\frac{4}{3})$，由切线定理得，两圆另一公切线l_2过点P，设

Analysis:

From the diagram, we can see that the two circles are externally tangent and both tangent to the line l_1 x = -1. The equation of the line l passing through the centers of the two circles $y=\frac{4}{3}x$.

The intersection point of this line and x = -1 is P(-1, -4/3). By the tangent theorem, the other common tangent line l_2 of the two circles passes through point P. Let

$l_2: y+\frac{4}{3}=k(x+1)$,

Distance Formula from Point to Line, we get:

$\frac{\left|k-\frac{4}{3}\right|}{\sqrt{k^2+1}}=1$ **then we get:** $k=\frac{7}{24}$ **which means:** $l_2: y=\frac{7}{24}x-\frac{25}{24}$

另由于两圆外切，因此在公切点处存在公切线 l_3 与 l 垂直，解得

"Since the two circles are externally tangent, there is a common tangent at the point of tangency that is perpendicular to the line joining their centers." So we get:

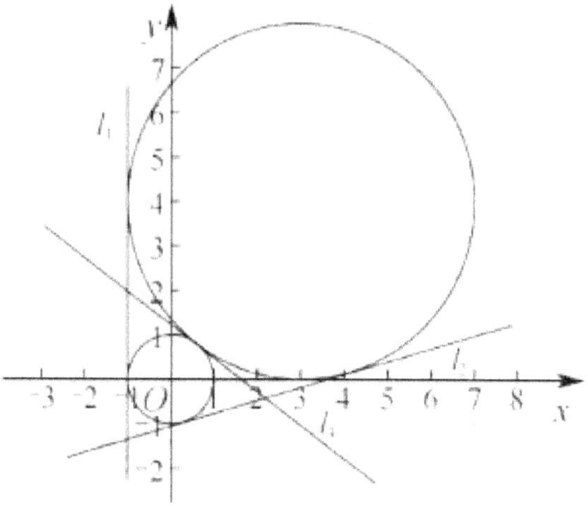

$l_3: y=-\frac{3}{4}x+\frac{5}{4}$

15. 若曲线 $y=(x+a)e^x$ 有两条过坐标原点的切线，则 a 的取值范围是________.

【答案】$(-\infty,-4)\cup(0,+\infty)$

Question 15:
If the curve y = (x+a)e^x has two tangent lines passing through the origin, what is the range of values for a?
Answer:
(-∞, -4) U (0, +∞)

【解析】易得曲线不过原点，设切点为 $\left(x_0, (x_0+a)e^{x_0}\right)$，则切线斜率为

Analysis:
It's easy to see that the curve doesn't pass through the origin. Let the tangent point $\left(x_0, (x_0+a)e^{x_0}\right)$, then the slope of the tangent line is... $f'(x_0)=(x_0+a+1)e^{x_0}$.

We get tangent function $y-(x_0+a)e^{x_0}=(x_0+a+1)e^{x_0}(x-x_0)$, and the tangent line pass the origin, we get $-(x_0+a)e^{x_0}=-x_0(x_0+a+1)e^{x_0}$, after simplification we get $x_0^2+ax_0-a=0$（※）, and we have two tangent lines, which is ※

方程有两不等实根，由判别式 $\Delta=a^2+4a>0$，得 $a<-4$，或 $a>0$.

The equation has two unequal real roots. This is determined by the discriminant

$\Delta = a^2 + 4a > 0$, we get: or $a < -4$, or $a > 0$.

16 已知椭圆$C: \frac{x^2}{a^2} + \frac{y^2}{b^2} = 1(a > b > 0)$，$C$的上顶点为$A$，两个焦点为$F_1$，$F_2$，离心率为$\frac{1}{2}$，过$F_1$且垂直于$AF_2$的直线与$C$交于$D$，$E$两点，$|DE| = 6$，则$\triangle ADE$的周长是________.

Question 16:
Given ellipse C: $x^2/a^2 + y^2/b^2 = 1$ (a > b > 0). Let A be the upper vertex of C, and F_1, F_2 be the two foci. The eccentricity is 1/2. A line pass through F_1 and perpendicular to A F_2 intersects C at points D and E. $|DE| = 6$, Find the perimeter of triangle DEF.
Answer: 13

【解析】椭圆离心率为$\frac{1}{2}$，不妨设$C: \frac{x^2}{4c^2} + \frac{y^2}{3c^2} = 1$，且$\triangle AF_1F_2$为正三角形，则直线$DE$斜率$k = \frac{\sqrt{3}}{3}$，由等腰三角形性质可得，$|AE| = |EF_2|$，$|AD| = |DF_2|$，由椭圆性质得$\triangle ADE$的周长等价于$|DE| + |DF_2| + |EF_2| = 4a$．另设直线$DE$方程为$y = \frac{\sqrt{3}}{3}(x + c)$，与椭圆方程联立得$13x^2 + 8cx - 32c^2 = 0$．

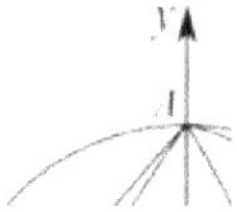

Analysis:

The ellipse's eccentricity is 1/2. We can assume the equation of the ellipse is $C: \frac{x^2}{4c^2} + \frac{y^2}{3c^2} = 1$, and Triangle $\triangle AF_1F_2$ is an equilateral triangle. The slope of line DE is $k = \frac{\sqrt{3}}{3}$. Due to the isosceles triangle properties, AE = EF, and the slope of DE is √3/3. AD = $|DF_2|$. Using the properties of the ellipse, we get:
The perimeter of triangle $\triangle ADE$ is equivalent to $|DE| + |DF_2| + |EF_2| = 4a$. Let the equation of line DE be $y = \frac{\sqrt{3}}{3}(x + c)$, and combine it with the equation of the ellipse to get $13x^2 + 8cx - 32c^2 = 0$.
Per Chord Length Formula $|DE| = \sqrt{k^2 + 1} \cdot |x_1 - x_2| = \sqrt{k^2 + 1} \cdot \sqrt{(x_1 + x_2)^2 - 4x_1x_2}$, we get: $|DE| = \sqrt{\frac{1}{3} + 1} \cdot \sqrt{(-\frac{8c}{13})^2 + \frac{128c^2}{13}} = \frac{48}{13}c = 6$, which means

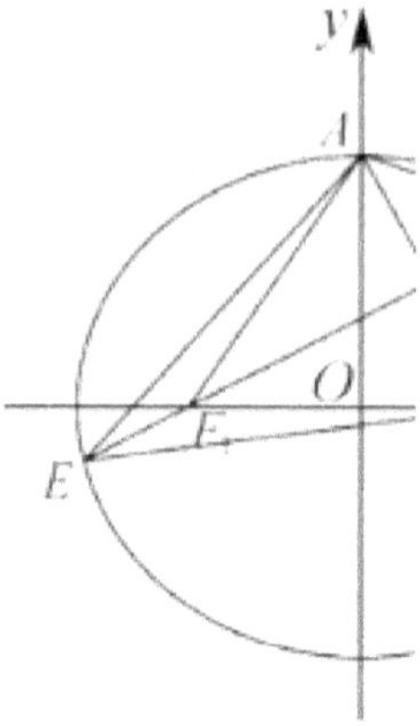

$c=\frac{13}{8}$， $4a=8c=13$.

四、解答题：本题共 6 小题，共 70 分。解答应写出文字说明、证明过程或演算步骤。

Section IV: Problem Solving
This section contains 6 questions for a total of 70 points. Answers must include explanations, proofs, or calculation steps.

17.（10 分）

记 S_n 为数列 $\{a_n\}$ 的前 n 项和，已知 $a_1=1$， $\{\frac{S_n}{a_n}\}$ 是公差为 $\frac{1}{3}$ 的等差数列.

（1）求 $\{a_n\}$ 得通项公式：

（2）证明： $\frac{1}{a_1}+\frac{1}{a_2}+\cdots+\frac{1}{a_n}<2$.

Question 17 (10 points):
Let S_n be the sum of the first n terms of the sequence $\{a_n\}$. Given that $a_1 = 1$ and $\{1/a_n\}$ is an arithmetic sequence with common difference 1/3.
(1) Find the general formula for $\{a_n\}$.
(2) Prove that $1/a_1 + 1/a_2 + ... + 1/a_n < 2$.

Analysis: （1） $S_1=a_1=1$， so $\frac{S_1}{a_1}=1$， so $\frac{S_n}{a_n}=1+(n-1)\cdot\frac{1}{3}=\frac{n+2}{3}$， so $S_n=\frac{n+2}{3}a_n$. when n>=2, $a_n=S_n-S_{n-1}=\frac{n+2}{3}a_n-\frac{n+1}{3}a_{n-1}$， so $(n-1)a_n=(n+1)a_{n-1}$， which means $\frac{a_n}{a_{n-1}}=\frac{n+1}{n-1}$ （$n\geqslant2$）； by Cumulative method, we get: $a_n=\frac{n(n+1)}{2}$ （$n\geqslant2$）， and $a_1=1$ Satisfies this formula, so the General formula is

$$(2)\ \frac{1}{a_1}+\frac{1}{a_2}+\cdots+\frac{1}{a_n}=2[\frac{1}{1\times 2}+\frac{1}{2\times 3}+\cdots+\frac{1}{n(n+1)}]$$

$$=2(1-\frac{1}{2}+\frac{1}{2}-\frac{1}{3}+\cdots+\frac{1}{n}-\frac{1}{n+1})$$

$$a_n=\frac{n(n+1)}{2}. \quad =2(1-\frac{1}{n+1})<2.$$

18.（12分）

记$\triangle ABC$的内角A，B，C的对边分别为a，b，c，已知$\frac{\cos A}{1+\sin A}=\frac{\sin 2B}{1+\cos 2B}$.

（1）若$C=\frac{2\pi}{3}$，求B；

（2）求$\frac{a^2+b^2}{c^2}$的最小值.

Question 18 (12 points):
Let triangle ABC have sides a, b, c opposite angles A, B, C respectively. Given:
cos A / (1 + sin A) = sin 2B / (1 + cos 2B)
(1) If C = 2π/3, find B.

(2) Find the minimum value of $\frac{a^2+b^2}{c^2}$.

Analysis From the conditions we get: $\sin 2B+\sin A\sin 2B=\cos A+\cos A\cos 2B$

$$\sin 2B=\cos A+\cos A\cos 2B-\sin A\sin 2B=\cos A+\cos(A+2B)$$

$$=\cos[\pi-(B+C)]+\cos[\pi-(B+C)+2B]$$

$$=-\cos(B+C)+\cos[\pi+(B-C)]$$

$$=-2\cos B\cos C$$

so

$$\sin B=-\cos C=\frac{1}{2},\quad B=\frac{\pi}{6}.$$

(2) from (1) we know $\sin B=-\cos C>0$, then

$$B=C-\frac{\pi}{2},\quad \sin B=\sin(C-\frac{\pi}{2})=-\cos C,$$

$$\sin A=\sin(B+C)=\sin(2C-\frac{\pi}{2})=-\cos 2C,$$

$$\frac{a^2+b^2}{c^2}=\frac{\sin^2 A+\sin^2 B}{\sin^2 C}=\frac{\cos^2 2C+\cos^2 C}{\sin^2 C}$$
$$=\frac{(1-2\sin^2 C)^2+(1-\sin^2 C)}{\sin^2 C}$$
$$=\frac{2+4\sin^4 C-5\sin^2 C}{\sin^2 C}=\frac{2}{\sin^2 C}+4\sin^2 C-5$$
$$\geq 2\sqrt{\frac{2}{\sin^2 C}\cdot 4\sin^2 C}-5=4\sqrt{2}-5,$$

By Law of Sines

And when $\sin^2 C=\frac{\sqrt{2}}{2}$ The equality sign holds so the minimum value of $\frac{a^2+b^2}{c^2}$ is $4\sqrt{2}-5$.

Question 19:

如图，直三棱柱 $ABC-A_1B_1C_1$ 的体积为4，$\triangle A_1BC$ 的面积为 $2\sqrt{2}$.

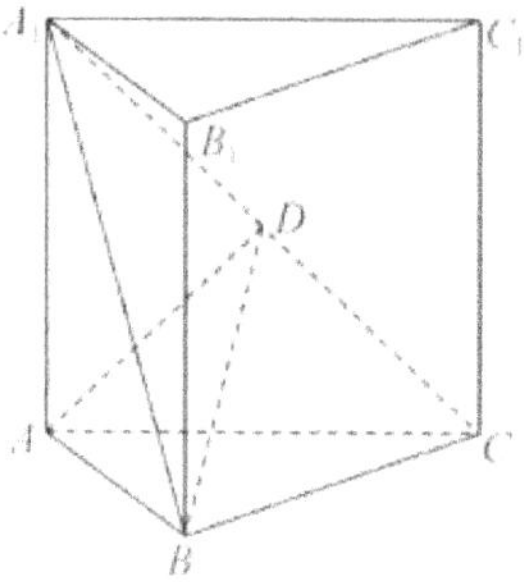

（1）求 A 到平面 A_1BC 的距离；

（2）设 D 为 A_1C 的中点，$AA_1=AB$，平面 $A_1BC\perp$ 平面 ABB_1A_1，求二面角 $A-BD-C$ 的正弦值.

Problem:

As shown in the figure, the volume of the right triangular prism $ABC-A_1B_1C_1$ is 4, and the area of triangle $\triangle A_1BC$ is $2\sqrt{2}$.

(1) Find the distance from point A to plane A_1BC.

(2) Let D be the midpoint of A_1C, and $AA_1=AB$. plane A_1BC ⊥ Plane ABB_1A_1. Find the sine of the dihedral angle A-BD-C.

Analysis:

Let the distance from point A to plane A_1BC is h,

$$V_{A_1-ABC}=\frac{1}{3}S_{\triangle ABC}\cdot A_1A=\frac{1}{3}V_{ABC-A_1B_1C_1}=\frac{1}{3}\times 4=\frac{4}{3},$$

$$V_{A-A_1BC}=\frac{1}{3}S_{\triangle A_1BC}\cdot h=\frac{1}{3}\times 2\sqrt{2}\cdot h,$$ so $\frac{1}{3}\times 2\sqrt{2}\cdot h=\frac{4}{3}$,

So $h=\sqrt{2}$, **Therefore, the distance from point A to plane A_1BC is √2.**

(2) 取 A_1B 的中点 E，连接 AE，

"Take point E as the midpoint of A_1B, and connect AE."

As $AA_1 = AB$, so $AE \perp A_1B$, as plane $A_1BC \perp$ plane $AE=\sqrt{2}$, then plane $A_1BC \cap$ plane $ABB_1A_1 = A_1B$, so $AE \perp$ plane A_1BC. $AE=\sqrt{2}$, then $AA_1 = AB=2$, so $AE \perp BC$, **Because it is a right triangular prism** $ABC - A_1B_1C_1$, **so** $A_1A \perp BC$, **because** $AE \cap A_1A = A$, **so BC** $\perp$ **plane** ABB_1A, **so** $BC \perp AB$, by

$V_{ABC-A_1B_1C_1} = \frac{1}{2}AB \cdot BC \cdot A_1A = \frac{1}{2} \times 2 \times BC \times 2 = 4$, so BC=2

English Translation:

Establish a three-dimensional rectangular coordinate system as shown in the figure, with BC as the x-axis, BA as the y-axis, and BB_1 **as the z-axis.**

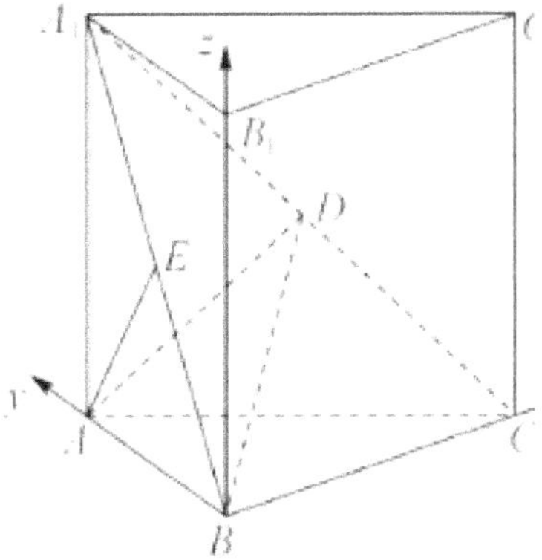

therefore:

$B(0,0,0)$, $A(0,2,0)$, $C(2,0,0)$, $A_1(0,2,2)$, $E(0,1,1)$, $D(1,1,1)$

Let Normal vector of plane BDC $\vec{n_1} = \vec{AE} = (0,-1,1)$, ,normal vector of plane BDA $\vec{n_2} = (x,y,z)$, $\vec{BA} = (0,2,0), \vec{BD} = (1,1,1)$, $\begin{cases} \vec{BA} \cdot \vec{n_2} = 0 \\ \vec{BD} \cdot \vec{n_2} = 0 \end{cases}$, so $\begin{cases} 2y = 0 \\ x + y + z = 0 \end{cases}$, so y=0

Let x=1, then z=-1, so $\vec{n_2} = (1,0,-1)$, so $\cos < \vec{n_1}, \vec{n_2} > = \frac{\vec{n_1} \cdot \vec{n_2}}{|\vec{n_1}| \cdot |\vec{n_2}|} = -\frac{1}{2}$,

Let 他 和 plane angler of Dihedral angle $A - BD - C$ as α then $\sin\alpha = \sqrt{1 - \cos\alpha} = \frac{\sqrt{3}}{2}$, so the sine of the dihedral angle A-BD-C is $\frac{\sqrt{3}}{2}$.

20.（12分）

一医疗团队为研究某地的一种地方性疾病与当地居民的卫生习惯（卫生习惯分为良好和不够良好两类）的关系，在已患该疾病的病例中随机调查了100例（称为病例组），同时在未患该疾病的人群中随机调查了100人（称为对照组），得到如下数据：

	不够良好	良好
病例组	40	60
对照组	10	90

（1）能否有99%的把握认为患该疾病群体与未患该疾病群体的卫生习惯有差异？

（2）从该地的人群中任选一人，A 表示事件"选到的人卫生习惯不够良好"，B 表示事件"选到的人患有该疾病"，$\frac{P(B|A)}{P(\overline{B}|A)}$ 与 $\frac{P(B|\overline{A})}{P(\overline{B}|\overline{A})}$ 的比值是卫生习惯不够良好对患该疾病风险程度的一项度量指标，记该指标为 R.

（i）证明：$R=\frac{P(A|B)}{P(\overline{A}|B)}\cdot\frac{P(\overline{A}|\overline{B})}{P(A|\overline{B})}$；

（ii）利用该调查数据，给出 $P(A|B)$，$P(A|\overline{B})$ 的估计值，并利用（i）的结果给出 R 的估计值.

附：$K^2=\frac{n(ad-bc)^2}{(a+b)(c+d)(a+c)(b+d)}$，

$P(K^2\geq k)$	0.050	0.010	0.001
k	3.841	6.635	10.828

Question 20 (12 points):

A medical team conducted a study on the relationship between a local disease in a certain area and the residents' hygiene habits (classified as good and not good). They randomly investigated 100 cases of people who had contracted the disease (called the case group) and 100 people who had not contracted the disease (called the control group), obtaining the following data:

	Not Good	Good
Case Group	40	60
Control Group	10	90

(1) Can it be concluded with 99% confidence that there is a difference in hygiene habits between the group with the disease and the group without the disease?

(2) Randomly select one person from the population of this area. Let A represent the event "the selected person has poor hygiene habits" and B represent the event "the selected person has the disease". The ratio of $\frac{P(B|A)}{P(\overline{B}|A)}$ 与 $\frac{P(B|\overline{A})}{P(\overline{B}|\overline{A})}$ is a measure of the degree of risk of poor hygiene habits for contracting the disease. Denote this index as R.

(i) Prove: $R=\dfrac{P(A|B)}{P(\overline{A}|B)}\cdot\dfrac{P(\overline{A}|\overline{B})}{P(A|\overline{B})}$;

(ii) Using the survey data, give estimates of $P(A|B)$, $P(A|\overline{B})$, and use the result of (i) to give an estimate of R.

Appendix: $K^2=\dfrac{n(ad-bc)^2}{(a+b)(c+d)(a+c)(b+d)}$,

$P(K^2\geq k)$	0.050	0.010	0.001
k	3.841	6.635	10.828

Analysis:

【解析】（1）假设患该疾病群体与未患该疾病群体的卫生习惯没有差异，

Analysis (1): Assume that there is no difference in hygiene habits between the group with the disease and the group without the disease.

$$K^2=\frac{200(40\times 90-60\times 10)^2}{50\times 150\times 100\times 100}=24>10.828,$$

Then:

所以有99%的把握认为患该疾病群体与未患该疾病群体的卫生习惯有差异；

Therefore, there is a 99% confidence that there is a difference in hygiene habits between the group with the disease and the group without the disease.

$$(2)（i）R=\frac{P(B|A)}{P(\overline{B}|A)}\cdot\frac{P(\overline{B}|\overline{A})}{P(B|\overline{A})}=\frac{\frac{P(AB)}{P(A)}}{\frac{P(A\overline{B})}{P(A)}}\cdot\frac{\frac{P(\overline{A}\,\overline{B})}{P(\overline{A})}}{\frac{P(\overline{A}B)}{P(\overline{A})}}=\frac{P(AB)}{P(A\overline{B})}\cdot\frac{P(\overline{A}\,\overline{B})}{P(\overline{A}B)}$$

$$=\frac{P(AB)}{P(\overline{A}B)}\cdot\frac{P(\overline{A}\,\overline{B})}{P(A\overline{B})}=\frac{\frac{P(AB)}{P(B)}}{\frac{P(\overline{A}B)}{P(B)}}\cdot\frac{\frac{P(\overline{A}\,\overline{B})}{P(\overline{B})}}{\frac{P(A\overline{B})}{P(\overline{B})}}=\frac{P(A|B)}{P(\overline{A}|B)}\cdot\frac{P(\overline{A}|\overline{B})}{P(A|\overline{B})},$$

We get:

$$P(A|B)=\frac{40}{100}=\frac{2}{5},\quad P(A|\overline{B})=\frac{10}{100}=\frac{1}{10},$$

(ii) from investment data we know

$$P(\overline{A}|B)=1-P(A|B)=\frac{3}{5},\quad P(\overline{A}|\overline{B})=\frac{9}{10},\quad 所以 R=\frac{\frac{2}{5}}{\frac{3}{5}}\cdot\frac{\frac{9}{10}}{\frac{1}{10}}=6.$$

Then

21.（12分）

已知点$A(2,1)$在双曲线$C:\frac{x^2}{a^2}-\frac{y^2}{a^2-1}=1(a>1)$上，直线$l$交$C$于$P$，$Q$两点，直线$AP$，$AQ$的斜率之和为0.

（1）求l的斜率；

（2）若$\tan\angle PAQ=2\sqrt{2}$，求$\triangle PAQ$的面积.

【答案】（1）l的斜率为0；（2）$\triangle PAQ$的面积为$\frac{16\sqrt{2}}{9}$

Question 21 (12 points):
Given that point A(2, 1) lies on the hyperbola C: $x^2/a^2 - y^2/(a^2-1) = 1$ ($a > 1$), and line l intersects C at points P and Q. The sum of the slopes of AP and AQ is 0.
(1) Find the slope of line l.
(2) If $\tan\angle PAQ = 2\sqrt{2}$, find the area of triangle PAQ.
Answer:
(1) The slope of line l is 0.
(2) The area of triangle PAQ is $16\sqrt{2} / 9$.
Analysis:
【解析】（1）将点A代入双曲线方程得 **Analysis (1):**
Substituting the point A(2, 1) into the hyperbola equation, we get:
$4/a^2 - 1/(a^2-1) = 1$
Simplify: $a^4-4a^2+4=0$, we get $a^2=2$, so the Hyperbola equation $\frac{x^2}{2}-y^2=1$;

由题显然直线l的斜率存在，设$l: y=kx+m$，设$P(x_1,y_1)Q(x_2,y_2)$，则联立直线与双曲线得：$(2k^2-1)x^2+4kmx+2m^2+2=0$，

It is obvious from the question that the slope of line l exists. Let y = kx + m, and let P(x_1, y_1) and Q(x_2, y_2). Then, combining the line and the hyperbola, we get:
$(2k^2-1)x^2+4kmx+2m^2+2=0$，

so

故$x_1+x_2=-\frac{4km}{2k^2-1}$，$x_1x_2=\frac{2m^2+2}{2k^2-1}$，

$$k_{AP}+k_{AQ}=\frac{y_1-1}{x_1-2}+\frac{y_2-1}{x_2-2}=\frac{kx_1+m-1}{x_1-2}+\frac{kx_2+m-1}{x_2-2}=0,$$

Simplify: $2kx_1x_2+(m-1-2k)(x_1+x_2)-4(m-1)=0$.

So $\frac{2k(2m^2+2)}{2k^2-1}+(m-1-2k)\left(-\frac{4km}{2k^2-1}\right)-4(m-1)=0$.

Which means:

$(k+1)(m+2k-1)=0$, and line l doesn't pass through point A, so k=-1.

Let tilt angle of line AP to be α ,by $\angle PAQ = 2\sqrt{2}$, we get $\tan\frac{\angle PAQ}{2}=\frac{\sqrt{2}}{2}$,

By $2\alpha+\angle PAQ=\pi$, we get $k_{AP}=\tan\alpha=\sqrt{2}$，即 $\frac{y_1-1}{x_1-2}=\sqrt{2}$.

Substitute the straight line l, we get m=5/3, so $x_1+x_2=\frac{20}{3}$，$x_1x_2=\frac{68}{9}$

And $|AP|=\sqrt{3}|x_1-2|$，$|AQ|=\sqrt{3}|x_2-2|$, by $\tan\angle PAQ=2\sqrt{2}$, we

get $\sin\angle PAQ=\frac{2\sqrt{2}}{3}$,

So

$$S_{\triangle PAQ}=\frac{1}{2}|AP||AQ|\sin\angle PAQ=\sqrt{2}\left|x_1x_2-2(x_1+x_2)+4\right|=\frac{16\sqrt{2}}{9}.$$

22.（12 分）

已知函数 $f(x)=e^x-ax$ 和 $g(x)=ax-\ln x$ 有相同的最小值.

（1）求 a；

（2）证明：存在直线 $y=b$，其与两条曲线 $y=f(x)$ 和 $y=g(x)$ 共有三个不同的交点，

并且从左到右的三个交点的横坐标成等差数列.

Question 22 (12 points):

Given functions $f(x)=e^x-ax$ and g(x) = ax - ln(x) have the same minimum value.
(1) Find a.
(2) Prove: There exists a line y = b that intersects the two curves y = f(x) and y = g(x) at three distinct points, and the x-coordinates of the three intersection points from left to right form an arithmetic sequence.
ANALYSIS:

【解析】（1）$f'(x)=e^x-a$，$g'(x)=a-\frac{1}{x}$

①$a\leqslant 0$时，$f'(x)>0$恒成立，所以$f(x)$在**R**上单调递增，即$f(x)$没有最小值.

该类情况应舍去.

②$a>0$时，$f'(x)$在$(-\infty,\ln a)$上小于0，在$(\ln a,+\infty)$上大于0.

Analysis (1):

$f'(x)=e^x-a$，$g'(x)=a-\dfrac{1}{x}$

① When a ≤ 0, f'(x) > 0 always holds, so f(x) is monotonically increasing on R, which means f(x) has no minimum value. This case should be discarded.

② When a > 0, f'(x) is less than 0 on (-∞, ln a) and greater than 0 on (ln a, +∞), ...

所以$f(x)$在$(-\infty,\ln a)$上单调递减，在$(\ln a,+\infty)$上单调递增，

所以$f(x)$在$x=\ln a$处有最小值为$f(\ln a)=a-a\ln a$，

所以$g'(x)$在$\left(0,\dfrac{1}{a}\right)$上小于0，在$\left(\dfrac{1}{a},+\infty\right)$上大于0，

所以$g(x)$在$\left(0,\dfrac{1}{a}\right)$上单调递减，在$\left(\dfrac{1}{a},+\infty\right)$上单调递增，

所以$g(x)$在$x=\dfrac{1}{a}$处有最小值为$g\left(\dfrac{1}{a}\right)=1+\ln a$，

Therefore, f(x) is decreasing on (-∞, ln a) and increasing on (ln a, +∞).
Therefore, f(x) has a minimum value at x = ln a, which is f(ln a) = a - a ln a.
Therefore, g'(x) is less than 0 on (0, 1) and greater than 0 on (1, +∞).
Therefore, g(x) is decreasing on (0, 1/a) and increasing on (1, +∞).
Therefore, g(x) has a minimum value at x = 1/a, which is g(1/a) = 1 + ln a.

因为$f(x)=e^x-ax$和$g(x)=ax-\ln x$有相同的最小值

所以有$f(\ln a)=a-a\ln a=g\left(\dfrac{1}{a}\right)=1+\ln a$，即$a-a\ln a=1+\ln a$

Because $f(x)=e^x-ax$ and g(x) = ax - ln(x) have the same minimum value,
So, f(ln a) = a - a ln a = g(1/a) = 1 + ln a, that is, a - a ln a = 1 + ln a.

因为$a>0$，所以上式等价于$\ln a-\dfrac{a-1}{a+1}=0$，

令$h(x)=\ln x-\dfrac{x-1}{x+1}(x>0)$，

Because a > 0, the above equation is equivalent to:

$$\ln a-\frac{a-1}{a+1}=0,$$

Let h $h(x)=\ln x-\frac{x-1}{x+1}(x>0)$.

则 $h'(x)=\frac{x^2+1}{x(x+1)^2}>0$ 恒成立，所以 $h(x)$ 在 $(0,+\infty)$ 上单调递增

又因为 $h(1)=0=h(a)$ 且 $a>0$，所以 $a=1$.

the $h'(x)=\frac{x^2+1}{x(x+1)^2}>0$ **always holds therefore, h(x) is monotonically increasing on (0, +∞).**
Also, because h(1) = 0 = h(a) and a > 0, we have a = 1.

(2) prove by $f(x)=e^x-x$, $g(x)=x-\ln x$, and f(x) is monotonically increasing on **(0, +∞). And g(x) is monotonically decreasing** on **(0, 1). And monotonically increasing** on **(1, +∞). And** $f(x)_{min}=g(x)_{min}=1$.

② when b=1, $f(x)_{min}=g(x)_{min}=1=b$, $y=b$ has 2 intersecting points with y=f(x) and y=g(x). the x-axis of the intersecting points are 1 and 1:

③ when b>1, first prove y=b and curve y=f(x) has two intersecting points:

which means $F(x)=f(x)-b$ has tow zero points,

$F'(x)=f'(x)=e^x-1$,

Therefore $F(x)$ is monotonically decreasing at $(-\infty,0)$,monotonically increasing at $(0,+\infty)$

And as

$F(-b)=e^{-b}>0$, $F(0)=1-b<0$, $F(b)=e^b-2b>0$,

Let

$t(b)=e^b-2b$，则 $t'(b)=e^b-2>0$, $t(b)>t(1)=e-2>0$

所以则 $F(x)=f(x)-b$ 在 $(-\infty,0)$ 上存在且只存在 1 个零点，设为 x_1，在 $(0,+\infty)$ 上存在且只存在 1 个零点，设为 x_2

Therefore, the function F(x) = f(x) - b has one and only one zero point on the interval (-∞, 0), denoted as x_1**, and one and only one zero point on the interval (0, +∞), denoted as** x_2**.**

其次，证明 $y=b$ 与曲线和有 2 个交点：

即证明 $G(x)=g(x)-b$ 有 2 个零点，$G'(x)=g'(x)=1-\frac{1}{x}$,

Secondly, prove that y = b intersects the curve at two points:

$$G'(x)=g'(x)=1-\frac{1}{x},$$

That is, prove that G(x) = g(x) - b has two zeros.

所以$G(x)$在$(0,1)$上单调递减，在$(1,+\infty)$上单调递增，

Therefore, G(x) is decreasing on the interval (0, 1) and increasing on the interval (1, +∞). And as

$$G\left(e^{-b}\right)=e^{-b}>0,\ G(1)=1-b<0,\ G(2b)=b-\ln 2b>0,$$

(let $\mu(b)=b-\ln 2b$, then

$$\mu'(b)=1-\frac{1}{b}>0,\ \mu(b)>\mu(1)=1-\ln 2>0$$

)

所以$F(x)=f(x)-b$在$(0,1)$上存在且只存在1个零点，设为x_3，在$(1,+\infty)$上存在且只存在1个零点，设为x_4。

Therefore, F(x) = f(x) - b has one and only one zero point on the interval (0, 1), denoted as x_3, and one and only one zero point on the interval (1, +∞), denoted as x_4.

再次，证明存在b使得$x_2=x_3$：

因为$F(x_2)=G(x_3)=0$，所以$b=e^{x_2}-x_2=x_3-\ln x_3$，

Secondly, prove that there exists b make x = x3 :

Because $F(x_2)=G(x_3)=0$, therefore $b=e^{x_2}-x_2=x_3-\ln x_3$.

If $x_2=x_3$, then $e^{x_2}-x_2=x_2-\ln x_2$, which means

$$e^{x_2}-2x_2+\ln x_2=0.$$

Therefore, we only need to prove there is solution for $e^x-2x+\ln x=0$ at (0,1)

Which means: there are zero point for $\varphi(x)=e^x-2x+\ln x$ 在$(0,1)$ at (0,1)

Because

$$\varphi\left(\frac{1}{e^3}\right)=e^{\frac{1}{e^3}}-\frac{2}{e^3}-3<0,\ \varphi(1)=e-2>0,$$

So $\varphi(x)=e^x-2x+\ln x$ has zero point at (0,1), take one zero point x_0, let $x_2=x_3=x_0$, that is $b=e^{x_0}-x_0$

则此时存在直线$y=b$，其与两条曲线$y=f(x)$和$y=g(x)$共有三个不同的交点

Then there exists a straight line y = b that intersects the two curves y = f(x) and y = g(x) at three distinct points.

最后证明$x_1+x_4=2x_0$，即从左到右的三个交点的横坐标成等差数列：

Finally, prove that $x_1+x_4=2x_0$, that is, the x-coordinates of the three intersection points from left to right form an arithmetic sequence.

because $F(x_1)=F(x_2)=F(x_0)=0=G(x_3)=G(x_0)=G(x_4)$,

so $F(x_1)=G(x_0)=F(\ln x_0)$,

又因为$F(x)$在$(-\infty,0)$上单调递减,

Because F(x) is strictly decreasing on the interval (-∞, 0),

$x_0 > 0$, that is $e^{x_0} > 1$, $x_1 > 1$, so $x_4 = e^{x_0}$,

Similarly, because $F(x_0) = G(e^{x_0}) = G(x_4)$,

又因为$G(x)$在$(1,+\infty)$上单调递增,

And because G(x) is monotonically increasing on (1, +∞).

$x_0 > 0$, that is $e^{x_0} > 1$, $x_1 > 1$, so $x_4 = e^{x_0}$,

And because $e^{x_0} - 2x_0 + \ln x_0 = 0$, so $x_1 + x_4 = e^{x_0} + \ln x_0 = 2x_0$.

即直线$y=b$,与两条曲线$y=f(x)$和$y=g(x)$从左到右的三个交点的横坐标成等差数列.

Therefore, the straight line y = b intersects the two curves y = f(x) and y = g(x) at three distinct points, and the x-coordinates of these three intersection points from left to right form an arithmetic sequence.

2024 年普通高等学校招生全国统一考试

全国甲卷理科数学

使用范围：陕西、宁夏、青海、内蒙古、四川

National Unified Entrance Examination for Ordinary Higher Education 2024
National Volume I, Mathematics
Applicable Regions: Shaanxi, Ningxia, Qinghai, Inner Mongolia, Sichuan

注意事项：

1．答题前，务必将自己的姓名、考籍号填写在答题卡规定的位置上．

2．答选择题时，必须使用 2B 铅笔将答题卡上对应题目的答案标号涂黑．如需改动，用橡皮擦擦干净后，再选涂其它答案标号．

3．答非选择题时，必须使用 0.5 毫米黑色签字笔，将答案书写在答题卡规定的位置上．

4．所有题目必须在答题卡上作答，在试题卷上答题无效．

5．考试结束后，只将答题卡交回．

Instructions:

1. Before answering, be sure to fill in your name and exam number in the designated areas on the answer sheet.
2. For multiple-choice questions, use a 2B pencil to fill in the corresponding answer on the answer sheet. If you need to change your answer, erase it completely before filling in a new one.
3. For non-multiple-choice questions, use a 0.5mm black pen to write your answers in the designated areas on the answer sheet.
4. All answers must be written on the answer sheet. Answers written on the question booklet will not be valid.
5. After the exam, only turn in the answer sheet.

一、选择题：本题共 12 小题，每小题 5 分，共 60 分．在每小题给出的四个选项中，只有一项是符合题目要求的．

Section I: Multiple Choice Questions

There are 12 questions in this section, with 5 points for each question, totaling 60 points. For each question, there are four options, only one of which is correct.

1． 设 $z=5+i$，则 $i(\bar{z}+z)=($ $)$

(A)10i (B)2i (C)10 (D)−2

【参考答案】A

【详细解析】因为 $z=5+i$，所以 $i(\bar{z}+z)=10i$，故选(A).

Question 1:

Let z = 5 + i. Then $i(\bar{z}+z)=($ $)$

(A) 10i (B) 2i (C) 10 (D) -2

Reference Answer: A

Detailed Explanation:

Because z = 5 + i, so $i(\bar{z}+z)=10i$, therefore the answer is (A).

2．( )集合 $A=\{1, 2, 3, 4, 5, 9\}$，$B=\{x|\sqrt{x}\subset A\}$，则 $\complement_A(A\cap B)=($ $)$

(A)$\{2, 3, 5\}$ (B)$\{2, 3, 5\}$ (C)$\{2, 3, 5\}$ (D)$\{2, 3, 5\}$

【参考答案】D

【详细解析】因为 $A=\{1, 2, 3, 4, 5, 9\}$，$B=\{x|\sqrt{x}\in A\}=\{1, 4, 9, 16, 25, 81\}$，所以 $\complement_A(A\cap B)=\{2, 3, 5\}$，故选(D).

Question 2:

Given set A = {1, 2, 3, 4, 5, 9} and set $A=\{1, 2, 3, 4, 5, 9\}$，$B=\{x|\sqrt{x}\subset A\}$，,
then A ∩ B =

(A)$\{2, 3, 5\}$ (B)$\{2, 3, 5\}$ (C)$\{2, 3, 5\}$ (D)$\{2, 3, 5\}$

Reference Answer: D

Detailed Explanation:

Because A = {1, 2, 3, 4, 5, 9} and $B=\{x|\sqrt{x}\in A\}=\{1, 4, 9, 16, 25, 81\}$，

therefore $\complement_A(A\cap B)=\{2, 3, 5\}$，So the answer is (D).

3. 若实数 x，y 满足约束条件(略)，则 $z=x-5y$ 的最小值为(　　)

(A)5　　(B)$\frac{1}{2}$　　(C)-2　　(D)$-\frac{7}{2}$

【参考答案】D

【详细解析】将约束条件两两联立可得 3 个交点：$(0,-1)$、$(\frac{3}{2},1)$和$(3,\frac{1}{2})$，经检验都符合约束条件．代入目标函数可得：$z_{min}=-\frac{7}{2}$，故选(D).

Question 3:
If real numbers x and y satisfy the constraint conditions (omitted), then the minimum value of z = x - 5y is ( ).
(A) 5 (B) ½ (C) -2 (D) -7/2
Reference Answer: D
Detailed Explanation:
By solving the constraint conditions pairwise, we obtain 3 intersection points: (0, -1), (3/2, 1), and (3,1/ 2). All of these points satisfy the constraint conditions. Substituting these points into the objective function, we get: $z_{min}=-\frac{7}{2}$, Therefore, the answer is (D).

4. 等差数列$\{a_n\}$的前 n 项和为 S_n，若 $S_5=S_{10}$，$a_5=1$，则 $a_1=$(　　)

(A)-2　　(B)$\frac{7}{3}$　　(C)1　　(D)2

【参考答案】B

【详细解析】因为 $S_5=S_{10}$，所以 $S_7=S_{18}$，$a_8=0$，又因为 $a_5=1$，所以公差 $d=-\frac{1}{3}$，$a_1=a_8-7d=\frac{7}{3}$，故选(B).

Question 4:

Let Sn be the sum of the first n terms of the arithmetic sequence {an}. If $S_5=S_{10}$, $a_5=1$, and $a_5=1$, then $a_1=($　　$)$
(A) -2 (B) 7/3 (C) 1 (D) 2
Reference Answer: B
Detailed Explanation:
Because $S_5=S_{10}$, so $S_7=S_{18}$, $a_8=0$, . Also, because $a_5=1$, so the common difference $d=-\frac{1}{3}$, $a_1=a_8-7d=\frac{7}{3}$, Therefore, the answer is (B).

5. 已知双曲线 C：$\frac{x^2}{a^2}-\frac{y^2}{b^2}=1(a>0, b>0)$的左、右焦点分别为 $F_1(0,$

(A) $\frac{13}{5}$ (B) $\frac{13}{7}$ (C) 2 (D) 3

【参考答案】C

【详细解析】$e=\frac{c}{a}=\frac{|F_1F_2|}{|PF_2|-|PF_1|}=2$，故选(C).

Question 5:
Given the hyperbola C: $x^2/a^2 - y^2/b^2 = 1$ ($a > 0, b > 0$), with left and right foci F1(0, c)

The eccentricity e of the hyperbola is:
(A) 13/5 (B) 13/7 (C) 2 (D) 3
Reference Answer: C
Detailed Explanation:
The eccentricity $e=\frac{c}{a}=\frac{|F_1F_2|}{|PF_2|-|PF_1|}=2$, Therefore, the answer is (C).

6. 曲线 $f(x)=x^6+3x$ 在(0，−1)处的切线与坐标轴围成的面积为(　　)

(A) $\frac{1}{6}$ (B) $\frac{\sqrt{3}}{2}$ (C) $\frac{1}{2}$ (D) $\frac{\sqrt{3}}{2}$

【参考答案】A

【详细解析】因为 $y'=6x^5+3$，所以 $k=3$，$y=3x-1$，$S=\frac{1}{2}\times\frac{1}{3}\times1=\frac{1}{6}$，故选(A).

Question 6:
The area enclosed by the tangent line to the curve $y = 6x^2 + 3x$ at point (0, -1) and the coordinate axes is:

(A) $\frac{1}{6}$ (B) $\frac{\sqrt{3}}{2}$ (C) $\frac{1}{2}$ (D) $\frac{\sqrt{3}}{2}$
Reference Answer: A
Detailed Explanation:

Because $y'=6x^5+3$, so $k=3$, $y=3x-1$, $S=\frac{1}{2}\times\frac{1}{3}\times1=\frac{1}{6}$.
Therefore, the answer is (A).

8. 已知 $\frac{\cos\alpha}{\cos\alpha-\sin\alpha}=\sqrt{3}$，则 $\tan(\alpha+\frac{\pi}{4})=$(　　)

(A) 3 (B) $2\sqrt{3}-1$ (C) -3 (D) $\frac{1}{3}$

【参考答案】B

【详细解析】因为 $\frac{\cos\alpha}{\cos\alpha-\sin\alpha}=\sqrt{3}$，所以 $\tan\alpha=1-\frac{\sqrt{3}}{3}$，$\tan(\alpha+\frac{\pi}{4})=\frac{\tan\alpha+1}{1-\tan\alpha}=2\sqrt{3}-1$，故选(B).

Question 8:

Given: $\frac{\cos\alpha}{\cos\alpha-\sin\alpha}=\sqrt{3}$, Find $\tan(\alpha+\frac{\pi}{4})=$(　　)

(A) 3 (B) $2\sqrt{3}-1$ (C) -3 (D)1/3

Reference Answer: B
Detailed Explanation:

Because $\frac{\cos\alpha}{\cos\alpha-\sin\alpha}=\sqrt{3}$,

therefore $\tan\alpha=1-\frac{\sqrt{3}}{3}$, $\tan(\alpha+\frac{\pi}{4})=\frac{\tan\alpha+1}{1-\tan\alpha}=2\sqrt{3}-1$,

Therefore, the answer is (B).

9. 已知向量 $a=(x+1, x)$，$b=(x, 2)$，则(　　)

(A) "$a\perp b$" 的必要条件是 "$x=-3$"　　(B) "$a // b$" 的必要条件是 "$x=-3$"

(C) "$a\perp b$" 的充分条件是 "$x=0$"　　(D) "$a // b$" 的充分条件是 "$x=0$"

【参考答案】C

【详细解析】$a\perp b$，则 $x(x+1)+2x=0$，解得：$x=0$ 或 -3，故选(C).

Question 9:
Given vectors a = (x+1, x) and b = (x, 2), then ( )

(A) "a ⊥ b" necessary condition is "x = -3"

(B) "$a // b$" necessary condition is "x = -3"

(C) "a ⊥ b" sufficient condition is "$x=0$"

(D) "$a // b$" sufficient condition is "x = 0"

Reference Answer: C
Detailed Explanation:
If a ⊥ b, then x(x+1) + 2x = 0, solving for x gives: x = 0 or -3, so the answer is (C).

10. 已知已知 m、n 是两条不同的直线，α、β 是两个不同的平面：①若 $m\perp\alpha$，$n\perp\alpha$，则 $m//n$；②若 $\alpha\cap\beta=m$，$m//n$，则 $n//\beta$；③若 $m//\alpha$，$n//\alpha$，m 与 n 可能异面，也可能相交，也可能平行；④若 $\alpha\cap\beta=m$，n 与 α 和 β 所成的角相等，则 $m\perp n$，以上命题是真命题的是(　　)

(A)①③　　(B)②③　　(C)①②③　　(D)①③④

【参考答案】A

【详细解析】选(A).

Question 10:
Given: m and n are two different lines, α and β are two different planes:

① If $m\perp\alpha$, $n\perp\alpha$, then m // n. ② If α ∩ β = m, m // n, then n // β. ③ If m // α and n // α, then m and n might be skew lines, intersecting lines, or parallel lines. ④ If α ∩ β = m, and the angles formed by n and α are equal to the angles formed by n and β, then $m\perp n$,
Which of the above statements are true?
(A) 1, 3 (B) 2, 3 (C) 1, 2, 3 (D) 1, 3, 4
Reference Answer: A
Detailed Explanation:

The correct answer is (A).

11. 在△ABC中，内角A，B，C所对边分别为a，b，c，若$B=\frac{\pi}{3}$，$b^2=\frac{9}{4}ac$，则$\sin A+\sin C=$(　　)

(A)$\frac{2\sqrt{39}}{13}$　　(B)$\frac{\sqrt{39}}{13}$　　(C)$\frac{\sqrt{7}}{2}$　　(D)$\frac{3\sqrt{13}}{13}$

【参考答案】C

【详细解析】因为$B=\frac{\pi}{3}$，$b^2=\frac{9}{4}ac$，所以$\sin A\sin C=\frac{4}{9}\sin^2 B=\frac{1}{3}$. 由余弦定理可得：$b^2=a^2+c^2$

Question 11:
In triangle ABC, sides a, b, and c are opposite to angles A, B, and C respectively. If B = π/3 $b^2=\frac{9}{4}ac$, , then sinA + sinC = ( )

(A)$\frac{2\sqrt{39}}{13}$　　(B)$\frac{\sqrt{39}}{13}$　　(C)$\frac{\sqrt{7}}{2}$　　(D)$\frac{3\sqrt{13}}{13}$

Reference Answer: C
Detailed Explanation:
Because $B=\frac{\pi}{3}$, $b^2=\frac{9}{4}ac$, we have $\sin A\sin C=\frac{4}{9}\sin^2 B=\frac{1}{3}$. Using the cosine rule, we can find $a^2 + c^2 = b^2$.

$-ac=\frac{9}{4}ac$, 即: $a^2+c^2=\frac{13}{4}ac$, $\sin^2 A+\sin^2 C=\frac{13}{4}\sin A\sin C=\frac{13}{12}$,

so $(\sin A+\sin C)^2=\sin^2 A+\sin^2 C$ $+2\sin A\sin C=\frac{7}{4}$, $\sin A+\sin C=\frac{\sqrt{7}}{2}$, Therefore, the answer is C

12.)已知a，b，c成等差数列，直线$ax+by+c=0$与圆C：$x^2+(y+2)^2=5$交于A，B两点，则$|AB|$的最小值为(　　)

(A)2　　(B)3　　(C)4　　(D)6

【参考答案】C

【详细解析】因为a，b，c成等差数列，所以$a-2b+c=0$，直线$ax+by+c=0$恒过$P(1,-2)$，当$PC\perp AB$时，$|AB|$取得最小值，此时$|PC|=1$，$|AB|=2\sqrt{5-|PC|^2}=4$，故选(C).

Question 12:
Given that a, b, and c form an arithmetic sequence, the line ax + by + c = 0 intersects the circle C: $x^2+(y+2)^2=5$ at points A and B. Find the minimum value of |AB|.
(A) 2 (B) 3 (C) 4 (D) 6
Reference Answer: C

Because a, b, and c form an arithmetic sequence, so a - 2b + c = 0, the line $ax+by+c=0$ always passes through point P(1, -2). When PC is perpendicular to AB, |AB| reaches its minimum value. At this point, $|AB|=2\sqrt{5-|PC|^2}=4$, . Therefore, the answer is (C).

13. 二项式$(\frac{1}{3}+x)^{10}$的展开式中系数的最大值是________.

【参考答案】5

【详细解析】展开式中系数最大的项一定在下面的 5 项：$C_{10}^{5}(\frac{1}{3})^5$、$C_{10}^{6}(\frac{1}{3})^4$、$C_{10}^{7}(\frac{1}{3})^3$、$C_{10}^{8}(\frac{1}{3})^2$、$C_{10}^{9}(\frac{1}{3})^1$，计算可得：系数的最大值为 $C_{10}^{8}(\frac{1}{3})^2=5$.

Question 13:

The maximum coefficient in the expansion of $(\frac{1}{3}+x)^{10}$ is ( )
Reference Answer: 5
Detailed Explanation:
The term with the maximum coefficient in the expansion must be one of the following five terms:

$C_{10}^{5}(\frac{1}{3})^5$、$C_{10}^{6}(\frac{1}{3})^4$、$C_{10}^{7}(\frac{1}{3})^3$、$C_{10}^{8}(\frac{1}{3})^2$、$C_{10}^{9}(\frac{1}{3})^1$,

Calculating these terms, we find the maximum coefficient is $C_{10}^{8}(\frac{1}{3})^2=5$.

14. 甲、乙两个圆台上下底面的半径均为 r_2 和 r_1，母线长分别为 $2(r_1-r_2)$和 $3(r_1-r_2)$，则两个圆台的体积之比$\frac{V_甲}{V_乙}$=________.

【参考答案】$\frac{\sqrt{6}}{4}$

【详细解析】$\frac{V_甲}{V_乙}=\frac{h_甲}{h_乙}=\frac{\sqrt{3}(r_1-r_2)}{2\sqrt{2}(r_1-r_2)}=\frac{\sqrt{6}}{4}$.

Question 14:

Two truncated cones have upper and lower base radii of r_2 和 r_1, . Their slant heights are $2(r_1-r_2)$和 $3(r_1-r_2)$, respectively. Find the ratio of the volumes of the two truncated cones: V_1/V_2.
Reference Answer: √6/4
Detailed Explanation:

$$\frac{V_甲}{V_乙}=\frac{h_甲}{h_乙}=\frac{\sqrt{3}(r_1-r_2)}{2\sqrt{2}(r_1-r_2)}=\frac{\sqrt{6}}{4}.$$

15. 已知 $a>1$，$\frac{1}{\log_8 a}-\frac{1}{\log_a 4}=-\frac{5}{2}$，则 $a=$________.

【参考答案】64

【详细解析】因为 $\frac{1}{\log_8 a}-\frac{1}{\log_a 4}=\frac{3}{\log_2 a}-\frac{1}{2}\log_2 a=-\frac{5}{2}$，所以 $(\log_2 a+1)(\log_2 a-6)=0$，而 $a>1$，故 $\log_2 a=6$，$a=64$.

Question 15:

Given a > 1, $\frac{1}{\log_8 a}-\frac{1}{\log_a 4}=-\frac{5}{2}$, Find the value of a.

Reference Answer: 64

Detailed Explanation:

Because $\frac{1}{\log_8 a}-\frac{1}{\log_a 4}=\frac{3}{\log_2 a}-\frac{1}{2}\log_2 a=-\frac{5}{2}$, have $(\log_2 a+1)(\log_2 a-6)=0$, Since a > 1, so $\log_2 a = 6$, so a = 64.

16. 编号为 1、2、3、4、5、6 的六个小球，不放回的抽取三次，记 m 表示前两个球号码的平均数，记 n 表示前三个球号码的平均数，则 m 与 n 差的绝对值不超过 0.5 的概率是________.

【参考答案】$\frac{7}{15}$

【详细解析】记前三个球的号码分别为 a、b、c，则共有 $A_6^3=120$ 种可能．令 $|m-n|=|\frac{a+b}{2}-\frac{a+b+c}{3}|=|\frac{a+b-2c}{6}|\leqslant 0.5$ 可得：$|a+b-2c|\leqslant 3$，根据对称性：$c=1$ 或 6 时，均有 2 种可能；$c=2$ 或 5 时，均有 10 种可能；$c=3$ 或 4 时，均有 16 种可能；故满足条件的共有 56 种可能，$P=\frac{56}{120}=\frac{7}{15}$.

Question 16:

There are six balls numbered 1, 2, 3, 4, 5, and 6. Three balls are drawn without replacement. Let m be the average of the numbers on the first two balls, and let n be the average of the numbers on all three balls. What is the probability that the absolute value of the difference between m and n is less than or equal to 0.5?

Reference Answer: 7/15

Detailed Explanation:

Let the numbers on the first three balls be a, b, and c. There are a total of A = 120 possible outcomes.

Let $|m-n|=|\frac{a+b}{2}-\frac{a+b+c}{3}|=|\frac{a+b-2c}{6}|\leqslant 0.5$

This simplifies to $|a+b-2c|\leqslant 3$, Due to symmetry, there are 2 possibilities when c = 1 or 6, 10 possibilities when c = 2 or 5, and 16 possibilities when c = 3 or 4. Therefore, there are a total of 56 possible outcomes that satisfy the condition. So the probability is 56/120 = 7/15.

三、解答题：共 70 分．解答应写出文字说明，证明过程或演算步骤．第 17 题～第 21 题为必考题，每个考题考生必须作答．第 22、23 题为选考题，考生根据要求作答．

Section III: Problem Solving
Total points: 70 points. Please write out your explanations, proofs, or calculation steps. Questions 17 to 21 are required for all students. Questions 22 and 23 are optional. Students should choose to answer based on requirements.

18．(12 分)(已知数列 $\{a_n\}$ 的前 n 项和为 S_n，且 $4S_n=3a_n+4$．

(1)求 $\{a_n\}$ 的通项公式；

(2)设 $b_n=(-1)^{n-1}na_n$，求数列 $\{b_n\}$ 的前 n 项和为 T_n．

【参考答案】见解析．

Question 18 (12 points)

Given that the sum of the first n terms of the sequence $\{a_n\}$ is S_n, and $4S_n=3a_n+4$.
(1) Find the general formula for the sequence $\{a_n\}$.
(2) Let $b_n=(-1)^{n-1}na_n$, Find the sum T_n of the first n terms of the sequence $\{b_n\}$.
[Reference Answer] See the solution.

【详细解析】(1)因为 $4S_n=3a_n+4$，所以 $4S_{n+1}=3a_{n+1}+4$，两式相减可得：$4a_{n+1}=3a_{n+1}-3a_n$，即：$a_{n+1}=-3a_n$，又因为 $4S_1=3a_1+4$，所以 $a_1=4$，故数列 $\{a_n\}$ 是首项为 4，公比为 -3 的等比数列，$a_n=4\cdot(-3)^{n-1}$；

Detailed Explanation:

(1) Because $4S_n=3a_n+4$, so $4S_{n+1}=3a_{n+1}+4$, subtracting the two equations gives: $4a_{n+1}=3a_{n+1}-3a_n$, That is $a_{n+1}=-3a_n$, and as $4S_1=3a_1+4$, so $a_1 = 4$, therefore the sequence $\{a_n\}$ is a geometric sequence with first term 4 and common ratio -3, $a_n = 4 * (-3)^{(n-1)}$.

(2)$b_n=(-1)^{n-1}na_n=4n\cdot 3^{n-1}$，所以 $T_n=4(1\cdot 3^0+2\cdot 3^1+3\cdot 3^2+\cdots+n\cdot 3^{n-1})$，$3T_n=4(1\cdot 3^1+2\cdot 3^2+3\cdot 3^3+\cdots+n\cdot 3^n)$，两式相减可得：$-2T_n=4(1+3^1+3^2+\cdots+3^{n-1}-n\cdot 3^n)=4(\frac{1-3^n}{1-3}-n\cdot 3^n)=(2-4n)3^n-2$，$T_n=(2n-1)3^n+1$．

(2) $b_n=(-1)^{n-1}na_n=4n\cdot 3^{n-1}$, so $T_n=4(1\cdot 3^0+2\cdot 3^1+3\cdot 3^2+\cdots+n\cdot 3^{n-1})$，$3T_n=4(1\cdot 3^1+2\cdot 3^2+3\cdot 3^3+\cdots+n\cdot 3^n)$, so the substruction of the two gives

$$-2T_n=4(1+3^1+3^2+\cdots+3^{n-1}-n\cdot 3^n)=4(\frac{1-3^n}{1-3}-n\cdot 3^n)=(2-4n)3^n-2，T_n=(2n-1)3^n+1.$$

解法 2：$b_n=(-1)^{n-1}na_n=4n\cdot3^{n-1}$，所以 $T_n=T_{n-1}+4n\cdot3^{n-1}$，两边同时减去$(2n-1)3^n$可得：$T_n-(2n-1)3^n=T_{n-1}-(2n-3)3^{n-1}$，故$\{T_n-(2n-1)3^n\}$为常数列，即：$T_n-(2n-1)3^n=1$，$T_n=(2n-1)3^n+1$.

Solution 2: $b_n=(-1)^{n-1}na_n=4n\cdot3^{n-1}$, so $T_n=T_{n-1}+4n\cdot3^{n-1}$. subtract $(2n-1)3^n$ on both sides gives $T_n-(2n-1)3^n=T_{n-1}-(2n-3)3^{n-1}$, so $\{T_n-(2n-1)3^n\}$ is a constant sequence, that is $T_n-(2n-1)3^n=1$，$T_n=(2n-1)3^n+1$.

19．(12分) 如图，已知 $AB\parallel CD$，$CD\parallel EF$，$AB=DE=EF=CF=2$，$CD=4$，$AD=BC=\sqrt{10}$，$AE=2\sqrt{3}$，M为CD的中点．

(1)证明：$EM\parallel$平面BCF；

(2)求二面角$A-EM-B$的正弦值．

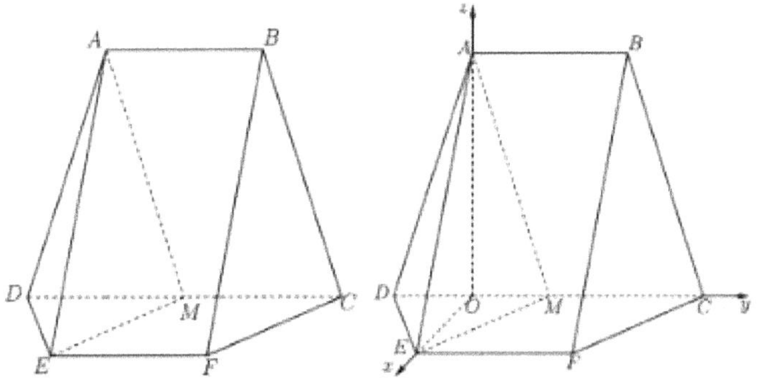

【参考答案】见解析

Given:

- AB is parallel to CD, CD is parallel to EF.
- AB = DE = EF = CF = 2
- CD = 4
- AD - BC = 10
- $AE=2\sqrt{3}$,
- M is the midpoint of CD

Question 19 (12 points):

(1) Prove that EM is parallel to plane BCF.

(2) Find the sine of the dihedral angle A-EM-B.

【详细解析】(1)由题意：$EF // CM$，$EF=CM$，而 $CF \subsetneqq$ 平面 ADO，$EM \not\subset$ 平面 ADO，所以 EM // 平面 BCF；

(2)取 DM 的中点 O，连结 OA，OE，则 $OA \perp DM$，$OE \perp DM$，$OA=3$，$OE=\sqrt{3}$，而 $AE=2\sqrt{3}$，

故 $OA \perp OE$．以 O 为坐标原点建立如图所示的空间直角坐标系，则 $A(0，0，3)$，$E(\sqrt{3}，0，0)$，$M(0，1，0)$，$B(0，2，3)$，$\overrightarrow{AE}=(\sqrt{3}，0，-3)$，$\overrightarrow{EM}=(-\sqrt{3}，1，0)$，$\overrightarrow{MB}=(0，1，3)$，

设平面 AEM 的法向量为 $\vec{n}=(x，y，z)$，由 $\begin{cases}\vec{n}\cdot\overrightarrow{AE}=0\\ \vec{n}\cdot\overrightarrow{EM}=0\end{cases}$ 可得：$\begin{cases}\sqrt{3}x-3z=0\\ -\sqrt{3}x+y=0\end{cases}$，令 $z=1$，则 $\vec{n}=(\sqrt{3}，$

$EF // CM$，$EF=CM$， and $CF \subsetneqq$ plane ADO, $EM \not\subset$ plane ADO, so EM// plane BCF;
(2) Let the midpoint of DM be O. Connect OA and OE. Then OA is perpendicular to DM, and OE is perpendicular to DM. OA = 3, OE = $\sqrt{3}$, , and AE = 2√3.
Therefore, OA is perpendicular to OE. Taking O as the origin, establish a spatial rectangular coordinate system as shown in the figure. Then $A(0，0，3)$，$E(\sqrt{3}，0，0)$，$M(0，1，0)$，$B(0，2，3)$，$\overrightarrow{AE}=(\sqrt{3}，0，-3)$，$\overrightarrow{EM}=(-\sqrt{3}，1，0)$，$\overrightarrow{MB}=(0，1，3)$，

Let the normal vector of plane AEM be $\vec{n}=(x，y，z)$， by $\begin{cases}\vec{n}\cdot\overrightarrow{AE}=0\\ \vec{n}\cdot\overrightarrow{EM}=0\end{cases}$,we get

$\begin{cases}\sqrt{3}x-3z=0\\ -\sqrt{3}x+y=0\end{cases}$ let z=1, then $\vec{n}=(\sqrt{3}，$

$3，1)$，同理：取平面 BEM 的法向量为 $\vec{m}=(\sqrt{3}，3，-1)$，则 $\cos<\vec{m}，\vec{n}>=\dfrac{\vec{m}\cdot\vec{n}}{|\vec{m}||\vec{n}|}=\dfrac{11}{13}$，$\sin<\vec{m}，\vec{n}>=\dfrac{4\sqrt{3}}{13}$，故二面角 $A-EM-B$ 的正弦值为 $\dfrac{4\sqrt{3}}{13}$．

Similarly, take the normal vector of plane BEM as $\vec{m}=(\sqrt{3}，3，-1)$， then $\cos<\vec{m}，\vec{n}>=\dfrac{\vec{m}\cdot\vec{n}}{|\vec{m}||\vec{n}|}=\dfrac{11}{13}$，$\sin<\vec{m}，\vec{n}>=\dfrac{4\sqrt{3}}{13}$， Therefore, the sine of the dihedral angle A-EM-B is $\dfrac{4\sqrt{3}}{13}$.

20．(12 分) 已知函数 $f(x)=(1-ax)\ln(1+x)-x$.

(1)当 $a=-2$ 时，求 $f(x)$的极值；

(2)当 $x\geq0$ 时，$f(x)\geq0$，求 a 的取值范围.

【参考答案】见解析

Problem 20 (12 points)
Given the function f(x) = (1 - ax)ln(1+x) - x.
(1) When a = -2, find the extreme values of f(x).
(2) When x ≥ 0, f(x) ≥ 0, find the range of values for a.

【详细解析】(1)当 $a=-2$ 时，$f(x)=(1+2x)\ln(1+x)-x$，$x>-1$．$f'(x)=2\ln(1+x)+\frac{x}{1+x}$，当 $x>0$ 时，$f'(x)>0$，当 $-1<x<0$ 时，$f'(x)<0$，所以 $f(x)$在$(-1，0)$上递减，在$(0，+\infty)$上递增，故 $f(x)$的极小值为 $f(0)=0$，无极大值；

Detailed Analysis:
When a = -2,

$$f(x)=(1+2x)\ln(1+x)-x,\ x>-1.\ f'(x)=2\ln(1+x)+\frac{x}{1+x},$$

When x > 0, f'(x) > 0, and when -1 < x < 0, f'(x) < 0. Therefore, f(x) is decreasing on (-1, 0) and increasing on (0, +∞). Hence, the minimum value of f(x) is f(0) = 0, and there is no maximum value.

(2)$f(x)=(1-ax)\ln(1+x)-x$，$f'(x)=-a\ln(1+x)-\frac{(a+1)x}{1+x}$．令 $g(x)=f'(x)$，则 $g'(x)=-\frac{a}{1+x}-\frac{a+1}{(1+x)^2}$．因为当 $x\geq0$ 时，$f(x)\geq0$，且 $f(0)=0$，$f'(0)=0$，所以 $g'(0)=-1-2a\geq0$，$a\leq-\frac{1}{2}$．
当 $a\leq-\frac{1}{2}$时，$g'(x)\geq\frac{1}{2(1+x)}-\frac{1}{2(1+x)^2}=\frac{x}{2(1+x)^2}\geq0$，$g(x)$在$[0，+\infty)$上递增，$g(x)=f'(x)\geq g(0)=0$，故 $f(x)$在$[0，+\infty)$上递增，$f(x)\geq f(0)=0$ 恒成立，即 a 的取值范围为$(-\infty，-\frac{1}{2}]$.

$$(2)f(x)=(1-ax)\ln(1+x)-x,\ f'(x)=-a\ln(1+x)-\frac{(a+1)x}{1+x}.$$

Let g(x) = f(x), then

$$g'(x)=-\frac{a}{1+x}-\frac{a+1}{(1+x)^2}.$$

Because when x ≥ 0, f(x) ≥ 0, and f(0) = 0, f'(0) = 0, so

$$g'(0)=-1-2a\geq0,\ a\leq-\frac{1}{2}.$$

When a ≤ -1/2,

$$g'(x)\geq\frac{1}{2(1+x)}-\frac{1}{2(1+x)^2}=\frac{x}{2(1+x)^2}\geq0,\ g(x)$$

g(x) is increasing on [0, +∞), g(x) = f'(x) > 0.
g(0) = 0, so f(x) is increasing on [0, +∞), f(x) = f(0) = 0 always holds, that is, the range of values for a is (-∞, -1/2].

21．(12分)(已知椭圆 C：$\frac{x^2}{a^2}+\frac{y^2}{b^2}=1(a>b>0)$的右焦点为 F，点 $M(1,\frac{3}{2})$在椭圆 C 上，且 $MF\perp x$ 轴.

(1)求椭圆 C 的方程；

(2)$P(4，0)$，过 P 的直线与椭圆 C 交于 A， B 两点，N 为 FP 的中点，直线 NB 与 MF 交于 Q，

证明：$AQ\perp y$ 轴.

【参考答案】见解析

Problem 21 (12 points)

Given the ellipse C: $x^2/a^2 + y^2/b^2 = 1$ ($a > b > 0$) with right focus F, and point M(1, 3/2) on the ellipse C, and MF is perpendicular to the x-axis.

(1) Find the equation of ellipse C.

(2)$P(4，0)$，

Let line l pass through point P and intersect ellipse C at points A and B. Let N be the midpoint of FP. line NB line intersects MF at Q. Prove that AQ is perpendicular to y-axis.

【详细解析】(1)设椭圆 C 的左焦点为 F_1，则$|F_1F|=2$，$|MF|=\frac{3}{2}$. 因为 $MF\perp x$ 轴，所以$|MF_1|=\frac{5}{2}$，$2a=|MF_1|+|MF|=4$，解得：$a^2=4$，$b^2=a^2-1=3$，故椭圆 C 的方程为：$\frac{x^2}{4}+\frac{y^2}{3}=1$；

(2)解法 1：设 $A(x_1，y_1)$，$B(x_2，y_2)$，$\overrightarrow{AP}=\lambda\overrightarrow{PB}$，则$\begin{cases}\frac{x_1+\lambda x_2}{1+\lambda}=4\\ \frac{y_1+\lambda y_2}{1+\lambda}=0\end{cases}$，即$\begin{cases}\lambda x_2=4+4\lambda-x_1\\ \lambda y_2=-y_1\end{cases}$. 又由

(1) Let ellipse C left focus to be F_1, the $|F_1F|=2，|MF|=\frac{3}{2}$.

Because $MF\perp x$ axis, so $|MF_1|=\frac{5}{2}，2a=|MF_1|+|MF|=4$,

We get: $a^2=4，b^2=a^2-1=3$, the function of ellipse C is $\frac{x^2}{4}+\frac{y^2}{3}=1$;

$\begin{cases}3x_1^2+4y_1^2=12\\ 3(\lambda x_2)^2+4(\lambda y_2)^2=12\lambda^2\end{cases}$可得：$3\cdot\frac{x_1+\lambda x_2}{1+\lambda}\cdot\frac{x_1-\lambda x_2}{1-\lambda}+4\cdot\frac{y_1+\lambda y_2}{1+\lambda}\cdot\frac{y_1-\lambda y_2}{1-\lambda}=12$，结合上式可得：$5\lambda-2\lambda x_2+3=0$. $P(4，0)$，$F(1，0)$，$N(\frac{5}{2}，0)$，则 $y_Q=\frac{3y_2}{5-2x_2}=\frac{3\lambda y_2}{5\lambda-2\lambda x_2}=-\lambda y_2=y_1$，故 $AQ\perp y$ 轴.

$\begin{cases}3x_1^2+4y_1^2=12\\ 3(\lambda x_2)^2+4(\lambda y_2)^2=12\lambda^2\end{cases}$

we get

$$3\cdot\frac{x_1+\lambda x_2}{1+\lambda}\cdot\frac{x_1-\lambda x_2}{1-\lambda}+4\cdot\frac{y_1+\lambda y_2}{1+\lambda}\cdot\frac{y_1-\lambda y_2}{1-\lambda}=12,$$

Combine the above formula we get:

$5\lambda-2\lambda x_2+3=0$. $P(4, 0)$, $F(1, 0)$, $N(\frac{5}{2}, 0)$,

Then $y_Q=\frac{3y_2}{5-2x_2}=\frac{3\lambda y_2}{5\lambda-2\lambda x_2}=-\lambda y_2=y_1$,

Therefore, $AQ\perp y$-axis

解法 2：设 $A(x_1, y_1)$，$B(x_2, y_2)$，，则 $\frac{y_1}{x_1-4}=\frac{y_2}{x_2-4}$，即：$x_1y_2-x_2y_1=4(y_2-y_1)$，所以 $(x_1y_2-x_2y_1)(x_1y_2+x_2y_1)=x_1^2y_2^2-x_2^2y_1^2=(4+\frac{4y_1^2}{3})y_2^2-(4+\frac{4y_2^2}{3})y_1^2=4(y_2-y_1)(y_2+y_1)=4(y_2-y_1)(x_1y_2+x_2y_1)$，

即：$x_1y_2+x_2y_1=y_2+y_1$，$2x_2y_1=5y_1-3y_2$．$P(4, 0)$，$F(1, 0)$，$N(\frac{5}{2}, 0)$，则 $y_Q=\frac{3y_2}{5-2x_2}=\frac{3y_1y_2}{5y_1-2y_1x_2}=y_1$，故 $AQ\perp y$ 轴．

Let $A(x_1, y_1)$, $B(x_2, y_2)$, , then $\frac{y_1}{x_1-4}=\frac{y_2}{x_2-4}$,

that is $x_1y_2-x_2y_1=4(y_2-y_1)$, Therefore, $(x_1y_2-x_2y_1)(x_1y_2+x_2y_1)=x_1^2y_2^2-x_2^2y_1^2=(4+\frac{4y_1^2}{3})y_2^2-(4+\frac{4y_2^2}{3})y_1^2=4(y_2-y_1)(y_2+y_1)=4(y_2-y_1)(x_1y_2+x_2y_1)$,

That is

$x_1y_2+x_2y_1=y_2+y_1$, $2x_2y_1=5y_1-3y_2$. $P(4, 0)$, $F(1, 0)$, $N(\frac{5}{2}, 0)$,

Then $y_Q=\frac{3y_2}{5-2x_2}=\frac{3y_1y_2}{5y_1-2y_1x_2}=y_1$, Therefore, $AQ\perp y$-axis

(二)选考题：共 10 分．请考生在第 22、23 题中任选一题作答，并用 2B 铅笔将所选题号涂黑，多涂、错涂、漏涂均不给分，如果多做，则按所做的第一题计分．

Section (II) Elective Questions: Total 10 points. Please choose one question from questions 22 and 23 to answer, and fill in the corresponding question number with a 2B pencil. Multiple choices, wrong choices, or missing choices will not be given points. If multiple questions are answered, only the first one will be graded.

22．[选修 4－4：坐标系与参数方程](10 分)(　　　　　　　　)在平面直角坐标系 xOy 中，以坐标原点 O 为极点，x 轴的正半轴为极轴建立极坐标系，曲线 C 的极坐标方程为$\rho=\rho\cos\theta+1$.

(1)写出 C 的直角坐标方程；

(2)直线$\begin{cases}x=t\\y=t+a\end{cases}$($t$ 为参数)与曲线 C 交于 A、B 两点，若$|AB|=2$，求 a 的值.

Question 22. (Elective 4-4: Coordinate Systems and Parametric Equations) (10 points)
In the rectangular coordinate plane xOy, with the origin as the pole and the positive x-axis as the polar axis, establish a polar coordinate system. The polar equation of curve C is p = pcosθ + 1.
(1) Write the rectangular coordinate equation of C.
(2) The line l: $\begin{cases}x=t\\y=t+a\end{cases}$ (where t is a parameter) intersects curve C at points A and B. If |AB| = 2, find the value of a.

【详细解析】(1)因为$\rho=\rho\cos\theta+1$，所以$\rho^2=(\rho\cos\theta+1)^2$，故 C 的直角坐标方程为：$x^2+y^2=(x+1)^2$，即：$y^2=2x+1$；

(2)将$\begin{cases}x=t\\y=t+a\end{cases}$代入 $y^2=2x+1$ 可得：$t^2+2(a-1)t+a^2-1=0$，$|AB|=\sqrt{2}|t_1-t_2|=\sqrt{16(1-a)}=2$，

解得：$a=\frac{3}{4}$.

)
Because p = pcosθ + 1, so p² = (pcosθ + 1)², therefore the rectangular coordinate equation of C is: x² + y² = (x + 1)², that is: y² = 2x + 1;
Detailed Analysis (2)
Substitute $\begin{cases}x=t\\y=t+a\end{cases}$ *into y² = 2x + 1, we get:*

$$t^2+2(a-1)t+a^2-1=0，|AB|=\sqrt{2}|t_1-t_2|=\sqrt{16(1-a)}=2，$$

So the solution: $a=\frac{3}{4}$.

23．[选修 4－5：不等式选讲](10 分)(　　　　　　　　)实数 a，b 满足 $a+b\geq3$.

(1)证明：$2a^2+2b^2>a+b$；

(2)证明：$|a-2b^2|+|b-2a^2|\geq6$.

[Elective 4-5: Inequality Selection] (10 points)
Real numbers a and b satisfy a + b ≥ 3.

(1) Prove: $2a^2 + 2b^2 > a + b$;
(2) Prove: $a - 2b^2 + 16 - 20^2 \geq 6$.

【解析】(1)因为 $a+b\geq 3$，所以 $2a^2+2b^2\geq (a+b)^2>a+b$;

(2)$|a-2b^2|+|b-2a^2|\geq |a-2b^2+b-2a^2|=|2a^2+2b^2-(a+b)|=2a^2+2b^2-(a+b)\geq (a+b)^2-(a$

$+b)=(a+b)(a+b-1)\geq 6$.

Analysis

(1) Solution: because $a+b\geq 3$, so $2a^2+2b^2\geq (a+b)^2>a+b$;
(2) **Solution:**

- $|a - 2b^2| + |b - 2a^2| \geq |(a - 2b^2) + (b - 2a^2)| = |2a^2 + 2b^2 - (a + b)|$
- Since $a + b \geq 3$, we know that $2a^2 + 2b^2 \geq (a + b)^2 > a + b$
- Therefore, $|a - 2b^2| + |b - 2a^2| \geq (a + b)^2 - (a + b) = (a + b)(a + b - 1) \geq 6$

Conclusion: $a - 2b^2 + 16 - 20^2 \geq 6$

2024全国高考新高考I卷

2024 National College Entrance Examination (New Curriculum I)

一.填空题（每题5分，共40分）

Section 1: Fill in the Blanks (5 points per question, total 40 points)

1. 已知集合$A=\{x|-5<x^3<5\}, B=\{-3,-1,0,2,3\}$，则$A\cap B=$

$A.\{-1,0\}$ $B.\{2,3\}$ $C.\{-3,-1,0\}$ $D.\{-1,0,2\}$

【答案】A

【解析】由$A=\{x|-5<x^3<5\}$得$x\in(-\sqrt[3]{5},\sqrt[3]{5})$所以$A\cap B=\{-1,0\}$选A

Given:

- Set A = {x | -5 < x < 5}
- Set B = {-3, -1, 0, 2, 3}

Find: A ∩ B (the intersection of sets A and B)

Options:

- A. (-1, 0)
- B. (2, 3)
- C. {-3, -1, 0}
- D. {-1, 0, 2}

Answer: A

Explanation: From set A, we know that x belongs to the interval (-5, √5). Therefore, the intersection of A and B is (-1, 0). So the answer is A.

2. 若$\frac{z}{z-1}=i+1$，则$z=$（ ）

$A.-1-i$ $B.-1+i$ $C.1-i$ $D.1+i$

【答案】C

【解析】$\frac{z}{z-1}=\frac{z-1+1}{z-1}=1+\frac{1}{z-1}=i+1$，$\frac{1}{z-1}=i$，$z=1-i$选C

2. 若$\frac{z}{z-1}=i+1$，

then z = ?

$A.-1-i$ $B.-1+i$ $C.1-i$ $D.1+i$

Answer: C

Explanation:

$$\frac{z}{z-1}=\frac{z-1+1}{z-1}=1+\frac{1}{z-1}=i+1,\ \frac{1}{z-1}=i,$$

Therefore, the answer is C.

3. 已知向量$\vec{a}=(0,1)$，$\vec{b}=(2,x)$若$\vec{b}\perp(\vec{b}-4\vec{a})$，则$x=$（ ）

$A.-2$ $B.-1$ $C.1$ $D.2$

【答案】D

【解析】$\vec{b}\perp(\vec{b}-4\vec{a})\Rightarrow x^2-4x+4=0\Rightarrow x=2$，选D

Given: $\vec{a}=(0,1)$，$\vec{b}=(2,x)$若$\vec{b}\perp(\vec{b}-4\vec{a})$，
find the value of x.

Options:

- A. -2
- B. -1
- C. 1
- D. 2

Answer: D

Explanation: $\vec{b}\perp(\vec{b}-4\vec{a})\Rightarrow x^2-4x+4=0\Rightarrow x=2$.

Therefore, x = 4. So the answer is D.

1. 已知$\cos(\alpha+\beta)=m,\tan\alpha\tan\beta=2$，则$\cos(\alpha-\beta)=$（　　）

$A.-3m$　　$B.-\frac{m}{3}$　　$C.\frac{m}{3}$　　$D.3m$

【答案】A

【解析】$\tan\alpha\tan\beta=2\Rightarrow\sin\alpha\sin\beta=2\cos\alpha\cos\beta$，$\cos(\alpha+\beta)=\cos\alpha\cos\beta-\sin\alpha\sin\beta=-\cos\alpha\cos\beta=m\Rightarrow\cos\alpha\cos\beta=-m$

$\cos(\alpha-\beta)=\cos\alpha\cos\beta+\sin\alpha\sin\beta=3\cos\alpha\cos\beta=-3m$，选A

Given $\cos(\alpha+\beta)=m,\tan\alpha\tan\beta=2$.

-

Find: cos(α - β)

Options:

- A. -3m
- B. -m/3
- C. m/3
- D. 3m

Answer: A

Explanation:

$\tan\alpha\tan\beta=2\Rightarrow\sin\alpha\sin\beta=2\cos\alpha\cos\beta$，$\cos(\alpha+\beta)=\cos\alpha\cos\beta-\sin\alpha\sin\beta=-\cos\alpha\cos\beta=$

$m\Rightarrow\cos\alpha\cos\beta=-m$

$\cos(\alpha-\beta)=\cos\alpha\cos\beta+\sin\alpha\sin\beta=3\cos\alpha\cos\beta=-3m$.

So the answer is A.

5. 已知圆柱和圆锥的底面半径相等，侧面积相等，且它们的高均为$\sqrt{3}$，则圆锥的体积为

$A.2\sqrt{3}\pi$　　$B.3\sqrt{3}\pi$　　$C.6\sqrt{3}\pi$　　$D.9\sqrt{3}\pi$

【答案】B

【解析】设圆柱和圆锥的底面半径为r高为$h=\sqrt{3}$　$\pi rl=2\pi rh\Rightarrow l=2h=2\sqrt{3}\Rightarrow r=\sqrt{l^2-h^2}=3$

$$V=\frac{1}{3}\pi r^2h=3\sqrt{3}\pi$$

选B

Given a cylinder and a cone with equal base radii and equal lateral surface areas. If their heights are both √3, find the volume of the cone.

Options:

- A. 2√3π
- B. 3√3π
- C. 6√3π
- D. 9√3π

Answer:

B. $3\sqrt{3}\pi$

Explanation:

Let the base radius of the cylinder and cone be r, and their height be h

$$h = \sqrt{3} \quad \pi r l = 2\pi r h \Rightarrow l = 2h = 2\sqrt{3} \Rightarrow r = \sqrt{l^2 - h^2} = 3$$

$$V = \frac{1}{3}\pi r^2 h = 3\sqrt{3}\pi$$

Therefore, the answer is B.

6. 已知函数$f(x) = \begin{cases} -x^2 - 2ax - a, x < 0 \\ e^x + ln(x+1), x \geq 0 \end{cases}$在$R$上单调递增，则 a 的取值范围是

$A(-\infty, 0]$ $B[-1,0]$ $C[-1,1]$ $D.[0,+\infty)$

【答案】B

【解析】

$f(x)$在R上单调递增，当$x \geq 0$单调递增，当$x < 0$时$-x^2 - 2ax - a$单调递增即可，所以

$$\begin{cases} -a \geq 0 \\ -a \leq 1 \end{cases} \Rightarrow a \in [-1,0]$$

Given the function $f(x) = \begin{cases} -x^2 - 2ax - a, x < 0 \\ e^x + ln(x+1), x \geq 0 \end{cases}$

If f(x) is monotonically increasing on R, find the range of values for a.

Options:

- A. (-∞, 0]
- B. [-1, 0]
- C. [-1, 1]
- D. [0, +∞)

Answer:

B. [-1, 0]

For f(x) to be monotonically increasing on R, when x>=0 monotonically increase, when x<0,

$x < 0$时$-x^2 - 2ax - a$ monotonically increase, so

$$\begin{cases} -a \geq 0 \\ -a \leq 1 \end{cases} \Rightarrow a \in [-1,0]$$

7. 当$x \in [0, 2\pi]$时，曲线$y = sin\,x$与$y = 2\,sin\left(3x - \frac{\pi}{6}\right)$的交点个数为

A. 3 B. 4 C. 6 D. 8

【答案】C

【解析】由图像可知有 6 个交点

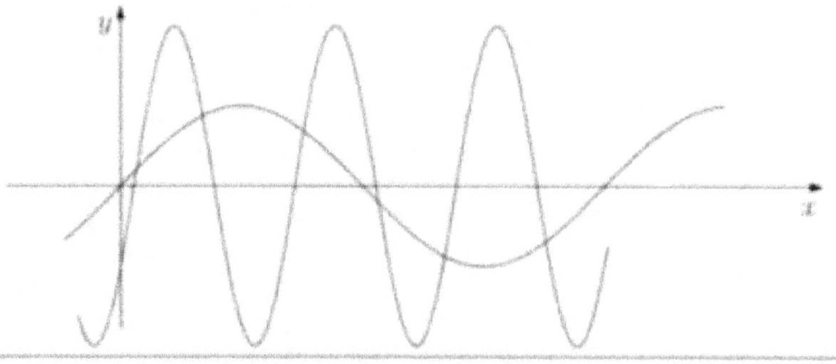

When $x \in [0,2\pi]$, how many intersection points do the curves y = sin x and $y = 2\sin\left(3x - \frac{\pi}{6}\right)$ have?

Options:

- A. 3
- B. 4
- C. 6
- D. 8

Answer:

C. 6

Explanation:

Based on the graph, there there are 6 intersection points.

8. 已知函数的定义域为R，$f(x) > f(x-1) + f(x-2)$且$x < 3$时$f(x) = x$，则下列结论中一定正确的是

A. $f(10) > 100$ B. $f(20) > 1000$ C. $f(10) < 1000$ D. $f(20) < 10000$

【答案】B

【解析】由题意知$x > 3$时$f(x)$可以趋近于正无穷，所以排除CD两项$f(1) = 1, f(2) = 2, f(x) > f(x-1) + f(x-2) \Rightarrow f(3) > 3, f(4) > 5, f(5) > 8, f(6) > 13$

$f(7) > 21, f(8) > 34, f(9) > 55, f(10) > 89$所以不一定满足$f(10) > 100$，所以选B

Given a function f(x) defined for all real numbers (R), such that:

- f(x) > f(x-1) + f(x-2) for x < 3
- f(x) = x for x< 3

Which of the following statements is definitely true?

- A. f(10) > 100
- B. f(20) > 1000
- C. f(10) < 1000
- D. f(20) < 10000

Answer:

B. f(20) > 1000

Explanation:

From the given conditions, we can see that f(x) grows rapidly to Positive Infinity for x > 3, so C&D can be eliminated. $f(1) = 1, f(2) = 2, f(x) > f(x-1) + f(x-2) \Rightarrow f(3) > 3, f(4) > 5, f(5) > 8, f(6) > 13$

$f(7) > 21, f(8) > 34, f(9) > 55, f(10) > 89$ cannot satisfy $f(10) > 100$.

Options A, C, and D can be eliminated based on this observation, so we select B.

二.多选题（每题6分，共18分）

Section II: Multiple Choice Questions (6 points per question, total 18 points)

9. 为了解推动出口后的亩收入（单位：万元）情况，从该种植区抽取样本，得到推动出口后亩收入的样本均值$\bar{X} = 2.1$，样本方差$S^2 = 0.01$，已知该种植区以往的亩收入X服从正态分布$N(1.8, 0.1^2)$，假设推动出口后的亩收入Y服从正态分布$N(\bar{X}, S^2)$，则（ ）（若随机变量Z服从正态分布$N(\mu, \sigma^2)$，则$P(Z < \mu + \sigma) \approx 0.8413$）

A. $P(X > 2) > 0.2$

B. $P(X > 2) < 0.5$

C. $P(Y > 2) > 0.5$

D. $P(Y > 2) < 0.8$

【答案】BC

【解析】

由题易知，$P(X > 1.9) \approx 0.1587$，A错B对；

$P(Y > 2) \approx 0.8413$，C对D错。

To understand the per-mu income (unit: 10,000 yuan) after promoting exports, a sample was

drawn from the planting area, and the sample mean of per-mu income after promoting exports, $\bar{X}$, was obtained.
The sample variance $S^2 = 0.01$. It is known that the previous per-mu income X of the planting area follows a normal distribution $N(1.8, 0.1^2)$. Assuming that the per-mu income Y after losing exports follows a normal distribution $N(\bar{X}, S^2)$, then (). (If random variable Z follows a normal distribution $N(\mu, \sigma^2)$, then $P(Z < \mu + \sigma) = 0.8413$)

- A. $P(X > 2) > 0.2$
- B. $P(X < 2) < 0.5$
- C. $P(Y > 2) > 0.5$
- D. $P(Y < 2) < 0.8$

Answer: BC
Explanation: From the graph, it is easy to know that $P(X > 1.9) = 0.1587$. A is wrong, B is correct:
$P(Y < 2) = 0.8413$. C is correct, D is wrong.
Summary:
The problem involves analyzing the income of a planting area before and after promoting exports. It uses statistical concepts like sample mean, variance, and normal distribution to calculate probabilities. The correct answers are B and C.

10. 设函数$f(x) = (x-1)^2(x-4)$，则（　　）
A. $x = 3$是$f(x)$的极小值点
B. 当$0 < x < 1$时，$f(x) < f(x^2)$
C. 当$1 < x < 2$时，$-4 < f(2x-1) < 0$
D. 当$-1 < x < 10$时，$f(2-x) > f(x)$

【答案】AC
【解析】
$f'(x) = 3(x-1)(x-3)$，$f(x)$在$(1,3)$递减，$(3,+\infty)$递增，则A对
$0 < x < 1$时，$x > x^2$，则$f(x) > f(x^2)$，B错
$1 < 2x-1 < 3$时，$f(3) < f(2x-1) < f(1)$，则$-4 < f(2x-1) < 0$，则C对
$f(2-x) > f(x)$可化为$(1-x)^2(-2-x) > (x-1)^2(x-4)$，可知$x = 1$时不成立，则D错

Given the function $f(x) = (x-1)^2(x-4)$, which of the following statements is true?

- A. $x = 3$ is a local minimum point of f(x).
- B. When $0 < x < 1$, $f(x) < f(x^2)$.
- C. When $1 < x < 2$, $-4 < f(2x-1) < 0$.
- D. When $-1 < x < 10$, $f(2-x) > f(x)$.

Answer:
AC (Both A and C are correct)
Explanation:

- By analyzing the first derivative f'(x), we find that f(x) decreases on the interval (1, 3) and increases on the interval $(3, +\infty)$. Therefore, $x = 3$ is a local minimum point, and statement A is correct.
- By comparing f(x) and $f(x^2)$ for $0 < x < 1$, we find that f(x) is actually greater than $f(x^2)$, so statement B is incorrect.
- By calculating the range of f(2x-1) for $1 < x < 2$, we find that it falls between -4 and 0, so statement C is correct.
- By comparing f(2-x) and f(x) for $-1 < x < 10$, we find a counterexample at $x = 1$, so statement D is incorrect.

Therefore, the correct answers are A and C.

Curve C can be seen as part of the curve in the figure. It is known that C passes through the origin O and the points on C satisfy the condition that the product of the distance from point F(2, 0) and the distance to the fixed line $x = a$ $(a < 0)$ is 4. Then ( ).

- A. a = -2
- B. Point $(2\sqrt{2}, 0)$ is on C
- C. The maximum y-coordinate of the points on C in the first quadrant is 1
- D. When point (x_0, y_0) is on C, $y_0 \le \frac{4}{x_0+2}$

11. 造型 可以看作图中的曲线C的一部分．已知C过坐标原点O．且C上的点满足横坐标大于−2．到点$F(2,0)$的距离与到定直线$x = a(a < 0)$的距离之积为4．则（　　）

A. $a = -2$

B. 点$(2\sqrt{2},0)$在C上

C. C在第一象限的点的纵坐标的最大值为1

D. 当点(x_0,y_0)在C上时．$y_0 \le \frac{4}{x_0+2}$

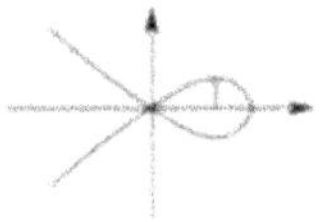

【答案】ABD

【解析】

由题意设C上一点$P(x,y)$．则满足$|x-a|\sqrt{(x-2)^2+y^2} = 4$

将(0,0)代入得．$|a| = 2$．即$a = -2$．A对．

且方程化为$(x+2)^2[(x-2)^2+y^2] = 16 \Rightarrow (x^2-4)^2 + y^2(x+2)^2 = 16$

将$(2\sqrt{2},0)$代入满足方程．则B对

$y^2 = \frac{x^2(8-x^2)}{(x+2)^2}$．对其整体求导得：$(y^2)' = -\frac{32}{(x+2)^2} - 2(x+2) + 8$

当$x = 2$时．$y = \pm 1$．但是$(y^2)'|_{x=2} < 0$．则在该点处函数保持单减趋势．则C错．

D可化简为$(x_0+2)\cdot\sqrt{(x_0-2)^2+y_0^2} = 4 \ge y_0\cdot(x_0+2)$

即$(x_0-2)^2 \ge 0$．显然成立．D对

Answer:
ABD (A, B, and D are correct)
Explanation:
Let P(x, y) be a point on C, then $|x-a|\sqrt{(x-2)^2+y^2} = 4$. Substituting (0, 0) into the equation, we get $|a| = 2$, so a = -2. A is correct.
The equation can be transformed into $(x+2)^2[(x-2)^2+y^2] = 16$, $(x^2-4)^2 + y^2(x+2)^2 = 16$. Substituting $(2\sqrt{2}, 0)$ into the equation, it satisfies the equation, so B is correct.

$y^2 = \frac{x^2(8-x^2)}{(x+2)^2}$, taking the derivative of the whole expression, we get: $(y^2)' = -\frac{32}{(x+2)^2} - 2(x+2) + 8$

When x = 2, y = ±1, but $(y^2)'|_{x=2} < 0$. Therefore, the function is decreasing at this point, so C is incorrect.

D can be simplified to $(x_0+2)\cdot\sqrt{(x_0-2)^2+y_0^2} = 4 \ge y_0\cdot(x_0+2)$
Since $(x-2)^2 \ge 0$, it is obviously true, so D is correct.

三.填空题（每题5分，共15分）

Section III: Fill in the Blanks (5 points per question, total 15 points)

12. 设双曲线$C: \frac{x^2}{a^2} - \frac{y^2}{b^2} = 1(a > 0, b > 0)$的左右焦点分别为$F_1, F_2$，过$F_2$与平行于y轴的直线交C于A,B两点，若$|F_1A| = 13, |AB| = 10$，则C的离心率为 ______

【答案】$\frac{3}{2}$

【解析】$|F_1A| - \frac{|AB|}{2} = 2a = 8 \Rightarrow a = 4$，同时$\frac{b^2}{a} = \frac{b^2}{4} = 5 \Rightarrow b^2 = 20, c^2 = 36$，从而$e = \frac{3}{2}$

Problem 12:
Let the left and right foci of the hyperbola C: $x^2/a^2 - y^2/b^2 = 1$ (a > 0, b > 0) be F_1 and F_2 respectively. A straight line parallel to the y-axis intersects the hyperbola C at points A and B. If $|F_1A| = 13$ and $|AB| = 10$, find the eccentricity of C.

Answer: 3/2

Analysis: $|F_1A| - \frac{|AB|}{2} = 2a = 8 \Rightarrow a = 4$, at the same time

$\frac{b^2}{a} = \frac{b^2}{4} = 5 \Rightarrow b^2 = 20, c^2 = 36$, so e = 3/2

13. 若曲线$y = e^x + x$在点(0,1)处的切线也是曲线$y = \ln(x+1) + a$的切线，则$a =$ ______

【答案】ln 2

【解析】设与$y = \ln(x+1) + a$的切点横坐标为x_0，则$k = f'(0) = 2 = \frac{1}{x_0+1} \Rightarrow x_0 = -\frac{1}{2}$，同时$y = e^x + x$在点(0,1)处的切线为$y = 2x + 1$，$y = \ln(x+1) + a$的切线为

$$y = 2x + 1 + a - \ln 2$$

则$1 + a - \ln 2 = 1 \Rightarrow a = \ln 2$

Problem 13:
If the tangent line to the curve y = e^x + x at point (0, 1) is also the tangent line to the curve y = ln(x+1) + a, then a =
Answer: ln 2
Explanation: Let the x-coordinate of the tangent point on y = ln(x+1) + a be x_0, then the slope k = f'(0) = 2 = 1/(x_0+1). Simultaneously, the tangent line to y = e^x + x at point (0, 1) is y = 2x + 1. The tangent line to y = ln(x+1) + a is y = 2x + 1 + a - ln 2. Therefore, 1 + a - ln 2 = 1, so a = ln 2.

14. 甲乙两人各有四张卡片，每张卡片上标有一个数字，甲的卡片上分别标有数字1,3,5,7，乙的卡片上分别标有数字2,4,6,8，两人进行四轮比赛，在每轮比赛中，两人各自从自己持有的卡片中随机选一张并比较所选卡片上数字的大小，数字大的人得1分，数字小的人得0分，然后各自弃置此轮所选的卡片（弃置的卡片在此后的轮次中不能使用），则四轮比赛后，甲的总得分小于2的概率为 ______

【答案】$\frac{1}{2}$

【解析】不妨设甲的顺序是1,3,5,7，考虑甲得分为0,1的情况

（1）0分情况：只有1种，$\begin{pmatrix}1&3&5&7\\2&4&6&8\end{pmatrix}$

（2）1分情况：

① 甲出3的时候得分，此时只有1种$\begin{pmatrix}1&3&5&7\\4&2&6&8\end{pmatrix}$

② 甲出5的时候得分，此时乙对应有两种情况

乙出4的时候有1种情况，乙出2的时候有2种情况，所以共3种；

③ 甲出7的时候得分，此时乙对应有3种情况

乙出6的时候有1种情况，

乙出4的时候有2种情况，

乙出2的时候有4种情况，从而共7种情况，

所以甲的总得分小于2的概率为

$$\frac{1+1+3+7}{A_4^4} = \frac{12}{24} = \frac{1}{2}$$

Problem 14:
There are two people, A and B, each with four cards. The numbers on A's cards are 1, 3, 5, and 7. The numbers on B's cards are 2, 4, 6, and 8. They play four rounds of a game. In each round, both players randomly select one card from their remaining cards and compare the numbers. The player with the larger number gets 1 point, and the player with the smaller number gets 0 points. The used cards are discarded after each round. What is the probability that A's total score

is less than 2 after four rounds?

Answer: 1/2

Explanation: Let's assume A's card order is 1, 3, 5, 7. Consider the cases where A's score is 0 or 1.

$$\begin{pmatrix}1\,3\,5\,7\\2\,4\,6\,8\end{pmatrix}$$

(1) 0 points: There is only one possibility:

(2) 1 point:

$$\begin{pmatrix}1\,3\,5\,7\\2\,4\,6\,8\end{pmatrix}$$

* A scores with 3: There is only one possibility:
* A scores with 5: there are two possibilities from B,
*B scores with 4: there is only one possibility
*B scores with 2: there are two possibilities, so a totle of 3 possibilities:
(1, 3, 5, 7), (2, 3, 5, 7), (2, 4, 5, 7)
* A scores with 7: There are seven possibilities for B:
(1, 3, 5, 7), (2, 3, 5, 7), (2, 4, 5, 7), (2, 4, 6, 7), (3, 4, 5, 7), (3, 4, 6, 7), (2, 3, 4, 7)
Therefore, the probability of A's total score being less than 2 is (1+1+3+7)/24 = 1/2.

四.解答题（第 15 题 13 分，第 16、17 题各 15 分，第 18、19 题各 17 分）

Section IV: Problem Solving (13 points for Question 15, 15 points each for Questions 16 and 17, 17 points each for Questions 18 and 19)

15. 记ΔABC的内角A,B,C的对边分别为a,b,c，已知$\sin C=\sqrt{2}\cos B$, $a^2+b^2-c^2=\sqrt{2}ab$

（1）求B；

（2）若ΔABC的面积$3+\sqrt{3}$，求c

【答案】（1）$\frac{\pi}{4}$；（2）$2\sqrt{2}$

【解析】1）$a^2+b^2-c^2=\sqrt{2}ab=2ab\cos C\Rightarrow C=\frac{\pi}{4}\Rightarrow \cos B=\frac{1}{2}\Rightarrow B=\frac{\pi}{3}$

（2）由（1）知$A=\frac{5}{12}\pi$，则

$$\frac{a}{\frac{\sqrt{6}+\sqrt{2}}{4}}=\frac{b}{\frac{\sqrt{3}}{2}}=\frac{c}{\frac{\sqrt{2}}{2}}$$

$$\Rightarrow a=\frac{\sqrt{3}+1}{2}c,\ b=\frac{\sqrt{6}}{2}c$$

$$S=\frac{1}{2}ab\sin C=\frac{1}{2}\cdot\frac{\sqrt{3}+1}{2}\cdot\frac{\sqrt{6}}{2}c^2\cdot\frac{\sqrt{2}}{2}=3+\sqrt{3}$$

所以$c=2\sqrt{2}$

Problem 15

Given triangle ABC with sides a, b, c opposite angles A, B, C respectively. It is known that:

$C=\sqrt{2}\cos B$, $a^2+b^2-c^2=\sqrt{2}ab$

(1) Find angle B.

(2) If the area of triangle ABC is 3 + √3, find c.

Answers: (1) B = π/4 (2) c = 2√2

Analysis: 1）$a^2+b^2-c^2=\sqrt{2}ab=2ab\cos C\Rightarrow C=\frac{\pi}{4}\Rightarrow \cos B=\frac{1}{2}\Rightarrow B=\frac{\pi}{3}$

2) from (1) we know $A = \frac{5}{12}\pi$, then

$$\frac{a}{\frac{\sqrt{6}+\sqrt{2}}{4}} = \frac{b}{\frac{\sqrt{3}}{2}} = \frac{c}{\frac{\sqrt{2}}{2}}$$

$$\Rightarrow a = \frac{\sqrt{3}+1}{2}c, b = \frac{\sqrt{6}}{2}c$$

$$S = \frac{1}{2}absinC = \frac{1}{2}\cdot\frac{\sqrt{3}+1}{2}\cdot\frac{\sqrt{6}}{2}c^2\cdot\frac{\sqrt{2}}{2} = 3+\sqrt{3}$$

Therefore: $c = 2\sqrt{2}$

16. 已知$A(0,3)$和$P\left(3,\frac{3}{2}\right)$为椭圆$C:\frac{x^2}{a^2}+\frac{y^2}{b^2}=1(a>b>0)$上两点

（1）求C的离心率；

（2）若过P的直线l交C于另一点B，且ΔABP的面积为9，求l的方程.

【答案】（1）$\frac{1}{2}$　　（2）$l: y=\frac{1}{2}x$或者$y=\frac{3}{2}x-3$

【解析】

（1）由题条件知

$$\begin{cases} b = 3 \\ \dfrac{9}{a^2} + \dfrac{9}{4b^2} = 1 \end{cases} \Rightarrow a^2 = 12, b^2 = 9$$

所以离心率为$\frac{1}{2}$

（2）

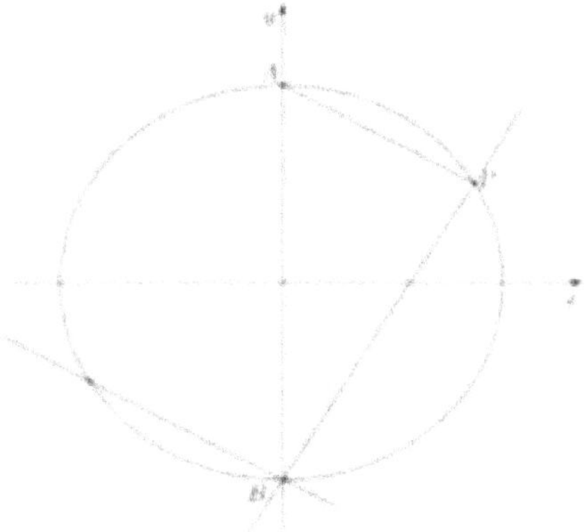

$$|PA| = \frac{3\sqrt{5}}{2}，PA: y = -\frac{1}{2}x + 3, x + 2y - 6 = 0$$

则设B在的直线$l': x + y + C = 0$，则

$$S = \frac{1}{2} \cdot \frac{3\sqrt{5}}{2} \cdot \frac{|6 + C|}{\sqrt{5}} = 9 \Rightarrow C = 6\ (-18舍)$$

所以$l': x + 2y + 6 = 0$。联立椭圆解得$B(0, -3)$或者$B\left(-3, -\frac{3}{2}\right)$，从而解得

$$l: y = \frac{1}{2}x 或者 y = \frac{3}{2}x - 3$$

Problem 16

Given points A(0, 3) and P(3, 3/2) on the ellipse C: $x^2/a^2 + y^2/b^2 = 1$ ($a > b > 0$).

(1) Find the eccentricity of the ellipse.

(2) If a line passing through P intersects the ellipse C at another point B, and the area of triangle ABP is 9, find the equation of the line.

Answers: (1) ½ (2) y = ½*x or y = 3/2*x – 3

(1) from the problem we know:

$$\begin{cases} b = 3 \\ \dfrac{9}{a^2} + \dfrac{9}{4b^2} = 1 \end{cases} \Rightarrow a^2 = 12, b^2 = 9$$

so the eccentricity of the ellipse is ½

(2)

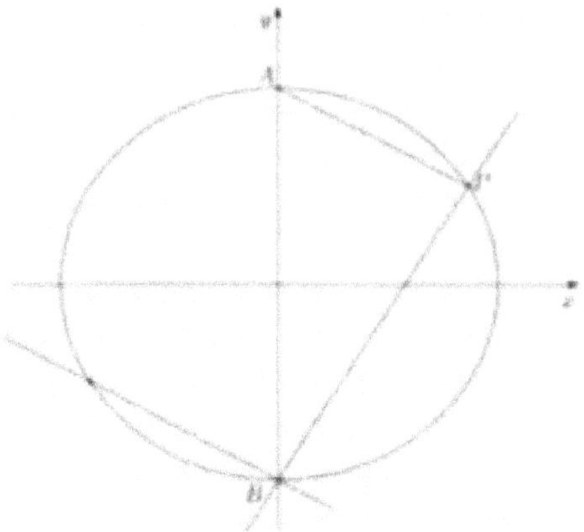

$|PA| = \frac{3\sqrt{5}}{2}$，$PA: y = -\frac{1}{2}x + 3, x + 2y - 6 = 0$

Then, let B on the line $l': x + y + C = 0$, then $S = \frac{1}{2} \cdot \frac{3\sqrt{5}}{2} \cdot \frac{|6+C|}{\sqrt{5}} = 9 \Rightarrow C = 6$（$-18$舍）
Therefore: l: x + 2y + 6 = 0. Solving the system of equations with the ellipse, we get B(0, -3) or B(-3, -3/2), thus obtaining l: y = ½*x or y = 3/2*x - 3

17. 如图：四棱锥$P-ABCD$中，$PA\perp$底面$ABCD$，$PA=AC=2$，$BC=1, AB=\sqrt{3}$

（1）若$AD\perp PB$，证明：$AD//$平面PBC；

（2）若$AD\perp DC$，且二面角$A-CP-D$的正弦值为$\frac{\sqrt{42}}{7}$，求AD

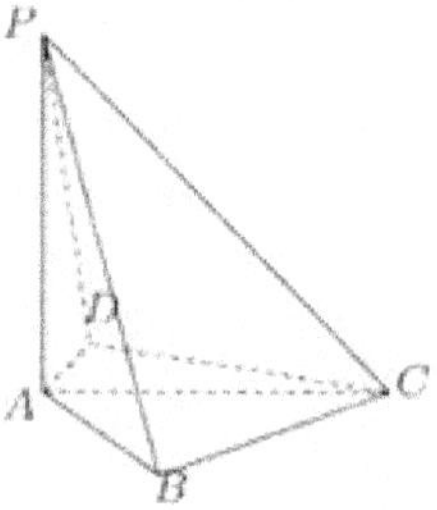

【答案】（1）略；（2）

【解析】

（1）证明：易知$\angle ABC=\frac{\pi}{2}$，因为$PA\perp AD, PB\perp AD \Rightarrow AD\perp PAB$，所以$AD\perp AB$，从而$AD//BC$，所以$AD//$平面$PBC$

（2）射影面积法

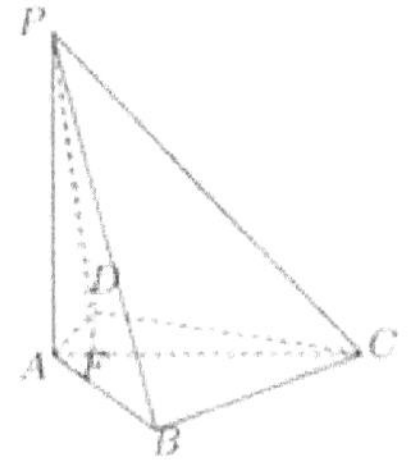

如图，作D在面PAC上的投影F，不妨设$AD = m$，则

$$\frac{CF}{\sqrt{4-m^2}} = \frac{\sqrt{4-m^2}}{2} \Rightarrow CF = \frac{4-m^2}{2}$$

则$A-CP-D$的夹角的余弦值

$$\cos\theta = \frac{\frac{1}{2}\cdot 2\cdot\frac{4-m^2}{2}}{\frac{1}{2}\sqrt{4-m^2}\cdot\sqrt{4+m^2}} = \frac{\sqrt{7}}{7}$$

$$\Rightarrow m = \sqrt{3}$$

Problem 17

As shown in the figure: In pyramid P-ABCD, PA is perpendicular to the base ABCD, PA = AC = 2. BC=1, AB = √3.

(1) If AD ⊥ PB, prove that AD is parallel to plane PBC.

(2) If AD ⊥ DC, and the sine of the dihedral angle A-CP-D is √42/7, find AD.

We know $\angle ABC = \frac{\pi}{2}$. Because PA ⊥ AD and PB ⊥ AD, AD ⊥ plane PAB. Therefore, AD ⊥ AB, so AD is parallel to BC. Therefore, AD is parallel to plane PBC.

(2) Projection area method

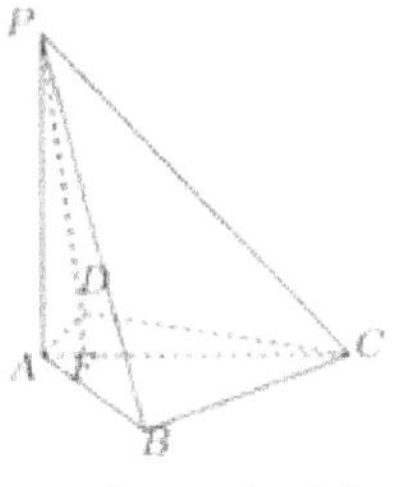

As shown in the figure, let F be the projection of point D onto plane PAC. Let AD = m. Then...

$$\frac{CF}{\sqrt{4-m^2}} = \frac{\sqrt{4-m^2}}{2} \Rightarrow CF = \frac{4-m^2}{2}$$

Then the cosine of the angle A-CP-D is...

$$\cos\theta = \frac{\frac{1}{2}\cdot 2\cdot\frac{4-m^2}{2}}{\frac{1}{2}\sqrt{4-m^2}\cdot\sqrt{4+m^2}} = \frac{\sqrt{7}}{7}$$

$$\Rightarrow m = \sqrt{3}$$

18. 已知函数$f(x)=ln\frac{x}{2-x}+ax+b(x-1)^3$

（1）若$b=0$，且$f'(x)\geq 0$，求a的最小值；

（2）证明：曲线$f(x)$为中心对称函数；

（3）若$f(x)>-2$，当且仅当$1<x<2$，求b的取值范围。

【答案】（1）$a=-2$　（3）$b\geq -\frac{2}{3}$

【解析】

（1）$f(x)=ln\frac{x}{2-x}+ax+b(x-1)^3$，$b=0$

$f(x)=ln\frac{x}{2-x}+ax$，$f'(x)=\frac{1}{x}+\frac{1}{2-x}+a\geq 0\Rightarrow a\geq -\frac{1}{x}-\frac{1}{2-x}$

$$-\frac{1}{x}-\frac{1}{2-x}=\frac{-2}{x(2-x)}\Rightarrow\left(-\frac{2}{x(2-x)}\right)_{max}=-2$$

故a的最小值为-2

（2）$f(x)+f(2-x)=ln\frac{x}{2-x}+ln\frac{2-x}{x}+ax+a(2-x)+b(x-1)^3+b(1-x)^3=2a$

故曲线$f(x)$关于$(1,a)$对称

（3）因为$f(1)=a\leq -2$，否则解集中含有$x=1$；又由（1）知$a\geq -2$，否则$f'(1)<0$，从而$a=-2$，$f(x)=\ln\frac{x}{2-x}-2x+b(x-1)^3$

$$f'(x)=\frac{2}{x(2-x)}-2+3b(x-1)^2=(x-1)^2\left[3b+\frac{2}{x(2-x)}\right]$$

设$g(x)=3b+\frac{2}{x(2-x)}$，$g(1)=3b+2$

正向：$b\geq -\frac{2}{3}$时，$f'(x)\geq 0$，$f(x)>f(1)=-2$

反向：$b<-\frac{2}{3}$时，由导数的保号性知存在$(1,\delta)$，$f'(x)<0$，此时$f(x)<-2$

所以$b\geq -\frac{2}{3}$

Problem 18

Given the function:

f(x) $f(x)=ln\frac{x}{2-x}+ax+b(x-1)^3$

(1) If b = 0 and f(x) ≥ 0, find the minimum value of a.

(2) Prove that the curve of f(x) is symmetric about a point.

(3) If f(x) > -2 holds true if and only if 1 < x < 2, find the range of values for b.

1）$a=-2$　（3）$b\geq -\frac{2}{3}$

Answer:

Analysis:

（1）$f(x)=ln\frac{x}{2-x}+ax+b(x-1)^3$，$b=0$

$f(x)=ln\frac{x}{2-x}+ax$，$f'(x)=\frac{1}{x}+\frac{1}{2-x}+a\geq 0\Rightarrow a\geq -\frac{1}{x}-\frac{1}{2-x}$

$$-\frac{1}{x}-\frac{1}{2-x}=\frac{-2}{x(2-x)}\Rightarrow\left(-\frac{2}{x(2-x)}\right)_{max}=-2$$

So the minimum value of a is -2.

（2）$f(x)+f(2-x)=ln\frac{x}{2-x}+ln\frac{2-x}{x}+ax+a(2-x)+b(x-1)^3+b(1-x)^3=2a$

So curve f(x) is symmetric about (1,a).

Because f(1) = a ≤ -2. Otherwise, the solution set would contain x = 1. Also, from (1) we know a ≥ -2. Otherwise, f'(1) < 0. Therefore, a = -2. f(x) = $\ln\frac{x}{2-x}$ - 2x + b(x-1)^3 /

$$f'(x) = \frac{2}{x(2-x)} - 2 + 3b(x-1)^2 = (x-1)^2\left[3b + \frac{2}{x(2-x)}\right]$$

Let $g(x) = 3b + \frac{2}{x(2-x)}$，$g(1) = 3b + 2$ Postive vector: when b>=-2/3, $f'(x) \geq 0$，$f(x) > f(1) = -2$

Negative vector:when $b < -\frac{2}{3}$,by the property of the sign of the derivative, there exists a number of x $(1,\delta)$, $f'(x) < 0$，此时$f(x) < -2$ therefore, b>=-2/3

19. 设m为正整数，数列$a_1, a_2, \ldots, a_{4m+2}$是公差不为0的等差数列．若从中删去两项$a_i$和$a_j (i < j)$后剩余的$4m$项可被平均分为$m$组，且每组的4个数都能构成等差数列，则称数列
$a_1, a_2, \ldots, a_{4m+2}$是$(i,j)$ 可分数列．
(1) 写出所有的$(i < j)$，$1 \leq i < j \leq 6$，使数列$a_1, a_2, \ldots, a_6$是$(i < j)$一可分数列；
(2) 当$m \geq 3$时，证明：数列$a_1, a_2, \ldots, a_{4m+2}$是$(2,13)$一可分数列；
(3) 从$1,2,\ldots,4m+2$中一次任取两个数i和$j(i < j)$记数列$a_1, a_2, \ldots, a_{4m+2}$是$(i < j)$一可分数列的概率为$P_m$，证明：

$P_m > \frac{1}{8}$.

【答案】(1) $(1,2),(5,6),(1,6)$；(2) 证明见解析；(3) 证明见解析

【解析】

(1) 略

(2) 该数列前14项中，a_1, a_4, a_7, a_{10}，a_3, a_6, a_9, a_{12}，a_5, a_8, a_{11}, a_{14}分别为等差数列．

则在去掉a_2, a_{13}后，该数列剩余的$4m + 2 - 14 = 4m - 12$项可被4整除，则后面连续4项均可构成等差数列，例如：$a_{15}, a_{16}, a_{17}, a_{18}; \ldots; a_{4m-1}, a_{4m}, a_{4m+1}, a_{4m+2}$分别为等差数列，证毕．

(3) 由 (1) 得：在a_{4m+2}整体数列中可取的项的下标应满足相邻关系或相隔$4k(k \geq 1)$项，则满足题意的(i,j)为：

	$i = 1$	$i = 5$	$i = 9$	…	$i = 4m - 3$
$j = 2$	$(1,2)$				
$j = 6$	$(1,6)$	$(5,6)$			
$j = 10$	$(1,10)$	$(5,6)$	$(9,10)$		
…	….	….	….	….	
$j = 4m + 2$	$(1,4m + 2)$	$(5,4m + 2)$	$(9,4m + 2)$	….	$(4m + 1,4m + 2)$

则符合题意的情况共$s_1 = (m + 1) + m + (m - 1) + \cdots + 1 = \frac{(m+1)(m+2)}{2}$种

由 (2) 得，取出a_2, a_9后，$a_1, a_3, a_5, a_7; a_4, a_6, a_8, a_{10}$分别为等差数列，则$a_{4m+2}$中剩余的$4m + 2 - 10 = 4m - 8$项中，相邻的四项均可分别构成等差数列，如：$a_{11}, a_{12}, a_{13}, a_{14}; \ldots; a_{4m-1}, a_{4m}, a_{4m+1}, a_{4m+2}$，则可推出满足题意的$(i,j)$为：

	$i=2$	$i=6$	$i=10$	$\cdots$	$i=4m-6$
$j=9$	$(2,9)$				
$j=13$	$(2,13)$	$(6,13)$			
$j=17$	$(2,17)$	$(6,17)$	$(10,17)$		
$\cdots$	$\cdots$	$\cdots$	$\cdots$	$\cdots$	
$j=4m+1$	$(2,4m+1)$	$(6,4m+1)$	$(10,4m+1)$	$\cdots$	$(4m-6,4m+1)$

则符合题意的情况共$s_2=(m-1)+(m-2)+\cdots+1=\frac{m(m-1)}{2}$种

综上以上两种情况，共$s_1+s_2=m^2+m+1$种，

则$P_m=\frac{m^2+m+1}{C_{4m+2}^2}=\frac{m^2+m+1}{8m^2+6m+1}=\frac{1}{8}\cdot\frac{8m^2+8m+8}{8m^2+6m+1}>\frac{1}{8}$，证毕.

Problem 19

Let m be a positive integer. The sequence $a_1, a_2, \ldots, a_{4m+2}$ is an arithmetic sequence with a non-zero common difference. If we delete two terms a_i and a_j ($i < j$) from the sequence, the remaining 4m terms can be divided into m groups, and the four numbers in each group form an arithmetic sequence. In this case, we call the sequence $a_1, a_2, \ldots, a_{4m+2}$ an (i, j)-divisible sequence.

(1) Find all pairs ($i< j$), where $1 \le i < j \le 6$, such that the sequence $a_1, a_2, \ldots, a_6$ is an (i,<j)-divisible sequence.

(2) Prove that when $m \ge 3$, the sequence $a_1, a_2, \ldots, a_{4m+2}$ is a (2, 13)-divisible sequence.

(3) Randomly select two different numbers i and j from 1, 2, ..., 4m+2 ($i < j$). Let Pm be the probability that the sequence $a_1, a_2, \ldots, a_{4m+2}$ is an (i< j)-divisible sequence. Prove that $P_m > \frac{1}{8}$.

Answer: (1) (1,2),(5,6),(1,6);

(2) Among the first 14 terms of the sequence, $a_1, a_2, a_3, a_4, a_5, a_6, a_7, a_8, a_9, a_{10}, a_{11}, a_{12}, a_{13}, a_{14}$ are respectively arithmetic sequences. Therefore, after removing a_2 and a_{13}, the remaining 4m+2-14 = 4m-12 terms are divisible by 4, so every consecutive four terms can form an arithmetic sequence, for example, $a_{m-1}, a_m, a_{m+1}, a_{m+2}$ are arithmetic sequences. Proven.

(3) From (1), the subscripts of the selectable terms in the entire sequence of 4m+2 should satisfy a neighboring relationship or be separated by 4k ($k \ge 1$) terms. Therefore, the (i, j) satisfying the conditions are:

	$i=1$	$i=5$	$i=9$	$\cdots$	$i=4m-3$
$j=2$	$(1,2)$				
$j=6$	$(1,6)$	$(5,6)$			
$j=10$	$(1,10)$	$(5,6)$	$(9,10)$		
$\cdots$	$\cdots$.	$\cdots$.	$\cdots$.	$\cdots$.	
$j=4m+2$	$(1,4m+2)$	$(5,4m+2)$	$(9,4m+2)$	$\cdots$.	$(4m+1,4m+2)$

则符合题意的情况共$s_1=(m+1)+m+(m-1)+\cdots+1=\frac{(m+1)(m+2)}{2}$种

由（2）得，取出a_2,a_9后，$a_1,a_3,a_5,a_7;a_4,a_6,a_8,a_{10}$分别为等差数列，则$a_{4m+2}$中剩余的$4m+2-10=4m-8$项中，相邻的四项均可分别构成等差数列，如：$a_{11},a_{12},a_{13},a_{14};\cdots;a_{4m-1},a_{4m},a_{4m+1},a_{4m+2}$，则可作出满足题意的$(i,j)$为：

Therefore,, the total number of cases that meet the conditions is

$s_1=(m+1)+m+(m-1)+\cdots+1=\frac{(m+1)(m+2)}{2}$种

From (2) we get: after taking out of a_2,a_9,, $a_1,a_3,a_5,a_7;a_4,a_6,a_8,a_{10}$ **as an arithmetic sequence, then the remaining** 4m+2 = 10 = 4m - 8.

	$i=2$	$i=6$	$i=10$	$\cdots$	$i=4m-6$
$j=9$	$(2,9)$				
$j=13$	$(2,13)$	$(6,13)$			
$j=17$	$(2,17)$	$(6,17)$	$(10,17)$		
$\cdots$	$\cdots$	$\cdots$	$\cdots$	$\cdots$	
$j=4m+1$	$(2,4m+1)$	$(6,4m+1)$	$(10,4m+1)$	$\cdots$	$(4m-6,4m+1)$

则符合题意的情况共$s_2=(m-1)+(m-2)+\cdots+1=\frac{m(m-1)}{2}$种

综上以上两种情况，共$s_1+s_2=m^2+m+1$种，

则$P_m=\frac{m^2+m+1}{C_{4m+2}^2}=\frac{m^2+m+1}{8m^2+6m+1}=\frac{1}{8}\cdot\frac{8m^2+8m+8}{8m^2+6m+1}>\frac{1}{8}$，证毕.

Therefore, the total number of cases that meet the conditions is

$s_2=(m-1)+(m-2)+\cdots+1=\frac{m(m-1)}{2}$，

Combining both cases, there are a total of $s_1+s_2=m^2+m+1$ possibilities.

Therefore, $P_m=\frac{m^2+m+1}{C_{4m+2}^2}=\frac{m^2+m+1}{8m^2+6m+1}=\frac{1}{8}\cdot\frac{8m^2+8m+8}{8m^2+6m+1}>\frac{1}{8}$, **proven.**

2024年普通高等学校招生全国统一考试

新高考数学II卷参考答案

本试卷共4页，19小题，满分150分，考试用时120分钟.

注意事项：

1. 答卷前，考生务必将自己的姓名、考生号、考场号和座位号填写在答题卡上，用2B铅笔将试卷类型(B)填涂在答题卡相应位置上，将条形码横贴在答题卡右上角“条形码粘贴处”.

2. 作答选择题时，选出每小题答案后，用2B铅笔在答题卡上对应题目选项的答案信息点涂黑；如需改动，用橡皮擦干净后，再选涂其他答案，答案不能答在试卷上.

3. 非选择题必须用黑色字迹的钢笔或签字笔作答，答案必须写在答题卡各题目指定区域内相应位置上；如需改动，先划掉原来的答案，然后再写上新答案；不准使用铅笔和涂改液. 不按以上要求作答无效.

4. 考生必须保持答题卡的整洁. 考试结束后，将试卷和答题卡一并交回.

一、单项选择题：本题共8小题，每小题5分，共40分. 在每小题给出的四个选项中，只有一项符合题目要求.

2024 National Unified Entrance Examination for Ordinary Higher Education
Mathematics II Answer Key
This paper consists of 4 pages, 19 questions, with a total score of 150 points and an examination time of 120 minutes.
Instructions:

1. **Before answering the questions, the candidate must fill in their name, candidate number, examination room number, and seat number on the answer sheet, and fill in the paper type (B) with a 2B pencil in the corresponding position on the answer sheet, and paste the bar code horizontally in the upper right corner of the answer sheet "Bar code pasting area".**
2. **When answering multiple-choice questions, after selecting the answer to each question, fill in the corresponding answer information point of the answer option on the answer sheet with a 2B pencil; if you need to change it, erase it with an eraser and then fill in other answers. The answer cannot be answered on the question paper.**
3. **Non-multiple-choice questions must be answered with a black ink pen or ballpoint pen, and the answers must be written in the corresponding position of the designated area for each question on the answer sheet; if you need to change it, cross out the original answer and then write the new answer; do not use pencils and correction fluid, and answers that do not comply with the above requirements will be invalid.**
4. **Candidates must keep the answer sheet clean. After the exam, return both the question paper and the answer sheet.**

Part I: Multiple Choice Questions
This part consists of 8 questions, 5 points for each question, with a total of 40 points. There is only one correct answer for each question among the four options given.

1. 已知 $z=-1-i$，则 $|z|=$ (　).

A: 0　　B: 1　　C: $\sqrt{2}$　　D: 2

答案：C.

解析：$|z|=\sqrt{(-1)^2+(-1)^2}=\sqrt{2}$. 故选C.

Given: z = -1 - i, find |z|.

- **A: 0**
- **B: 1**
- **C: $\sqrt{2}$**
- **D: 2**

Answer: C.

Explanation: $|z| = \sqrt{(-1)^2 + (-1)^2} = \sqrt{2}$. **Therefore, the answer is C.**

2. 已知命题 $p: \forall x \in \mathbf{R}, |x+1| > 1$; 命题 $q: \exists x > 0, x^3 = x$. 则（ ）.

A: p 和 q 都是真命题

B: $\neg p$ 和 q 都是真命题

C: p 和 $\neg q$ 都是真命题

D: $\neg p$ 和 $\neg q$ 都是真命题

答案：B.

解析：$p: \forall x \in \mathbf{R}, |x+1| > 1$, 假, 则 $\neg p$ 为真; $q: \exists x > 0, x^3 = x$, 真, 则 $\neg q$ 为假. 故选 B.

Given: Proposition p: For $\forall x \in \mathbf{R}, |x+1| > 1$; **Proposition q:** $q: \exists x > 0, x^3 = x$. **Then ().**

- **A: Both p and q are true propositions.**
- **B: Both -p and q are true propositions.**
- **C: p is a true proposition and q is a false proposition.**
- **D: p is a false proposition and q is a true proposition.**

Answer: B.

Explanation: $p: \forall x \in \mathbf{R}, |x+1| > 1$, **is false, so its negation is true. q:** $\exists x > 0, x^3 = x$, **is true, then -q is false. Therefore, the answer is B.**

3. 已知向量 a, b 满足 $|a| = 1, |a + 2b| = 2$, 且 $(b - 2a) \perp b$. 则 $|b| =$（ ）.

A: $\frac{1}{2}$

B: $\frac{\sqrt{2}}{2}$

C: $\frac{\sqrt{3}}{2}$

D: 1

答案：B.

解析：$|a| = 1, |a + 2b| = 2$.

$$(b - 2a) \perp b \Rightarrow b^2 - 2a \cdot b = 0, \quad ①$$

$$|a + 2b| = 2 \Rightarrow a^2 + 4a \cdot b + 4b^2 = 4. \quad ②$$

联立 ①② 得 $|b| = \frac{\sqrt{2}}{2}$. 故选 B.

Given: Vectors a and b satisfy |a| = 1, |a + 2b| = 2, and (b - 2a) $\perp$ **b . Find |b|.**

- **A: 1/2**
- **B: √2/2**
- **C: √3/2**
- **D: 1**

Answer: B.

Solution: From |a| = 1 and |a + 2b| = 2, we get:

$$(b - 2a) \perp b \Rightarrow b^2 - 2a \cdot b = 0, \quad ①$$

$$|a + 2b| = 2 \Rightarrow a^2 + 4a \cdot b + 4b^2 = 4. \quad ②$$

Solve the system of equations ① and ② to get $b^2 = 1/2$. Therefore, choose B.

5. 已知曲线 $C: x^2+y^2=16\ (y>0)$，从 C 上任意一点 P 向 x 轴作垂线段 PP'，P' 为垂足，则线段 PP' 的中点 M 的轨迹方程为（　）.

A: $\frac{x^2}{16}+\frac{y^2}{4}=1\ (y>0)$　　B: $\frac{x^2}{16}+\frac{y^2}{8}=1\ (y>0)$

C: $\frac{y^2}{16}+\frac{x^2}{4}=1\ (y>0)$　　D: $\frac{y^2}{16}+\frac{x^2}{8}=1\ (y>0)$

答案：A.

解析：如图，$x^2+y^2=16\ (y>0)$.

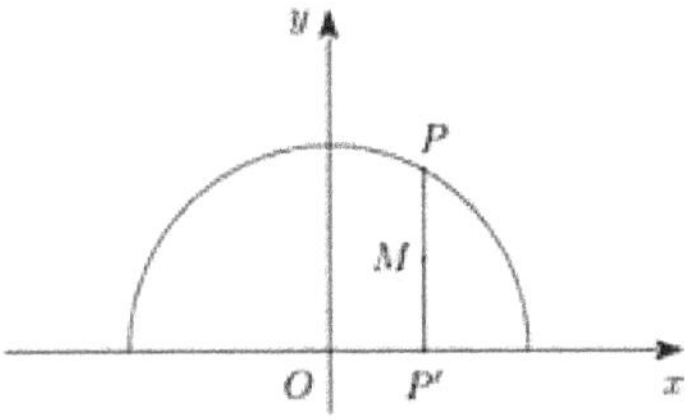

设 $M(x_0,y_0)\ (y_0>0)$，则 $P'(x_0,0)$，$P(x_0,2y_0)$，代入 $x^2+y^2=16 \Rightarrow x_0^2+4y_0^2=16\ (y_0>0)$.

易得，M 的轨迹方程为 $\frac{x^2}{16}+\frac{y^2}{4}=1\ (y>0)$，故选 A.

Given: Curve C: $x^2 + y^2 = 16$ (p > 0). From any point P on C, draw a perpendicular line PP' to the x-axis, with P' being the foot of the perpendicular. Find the equation of the locus of the midpoint M of line segment PP'.

A: $\frac{x^2}{16}+\frac{y^2}{4}=1\ (y>0)$　　B: $\frac{x^2}{16}+\frac{y^2}{8}=1\ (y>0)$

C: $\frac{y^2}{16}+\frac{x^2}{4}=1\ (y>0)$　　D: $\frac{y^2}{16}+\frac{x^2}{8}=1\ (y>0)$

Answer A.

Analysis: as figure:

$x^2+y^2=16\ (y>0)$.

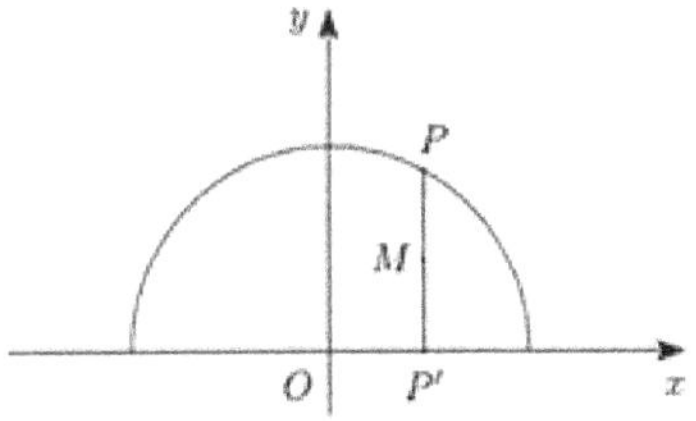

Let $M(x_0,y_0)\ (y_0>0)$, **then** $P'(x_0,0)$, $P(x_0,2y_0)$, 代入 $x^2+y^2=16 \Rightarrow x_0^2+4y_0^2=16\ (y_0>0)$.

It is easy to obtain that the equation of the locus of M is $x^2/16 + y^2/4 = 1$ (y > 0). Therefore, choose A.

6. 设函数 $f(x)=a(x+1)^2-1$，$g(x)=\cos x+2ax$（a 为常数），当 $x\in(-1,1)$ 时，曲线 $y=f(x)$ 与 $y=g(x)$ 恰有一个交点，则 $a=$（　）.

A: -1　　B: $\frac{1}{2}$　　C: 1　　D: 2

答案：D.

解析：令 $f(x)=g(x)$，则 $\cos x=a(x^2+1)-1$.

令 $h(x)=\cos x-a(x^2+1)+1$. 因为 $h(x)$ 为偶函数，且 $h(x)$ 有唯一零点，所以有 $h(0)=0$，即 $a=2$. 故选 D.

Prompt: Let $f(x)=a(x+1)^2-1, g(x)=\cos x+2ax$ (a is constant) When $x\in(-1,1)$ the curves y = f(x) and y = g(x) have exactly one intersection point. Find the value of a .

Answer D

Analysis: Let f(x) = g(x), then cos(x) = $a(x^2 + 1) - 1$.

Let h(x) = cos(x) - $a(x^2 + 1) + 1$. Because h(x) is an even function and h(x) has a unique zero point, we have h(0) = 0, that is, a = 2. Therefore, choose D.

7. 已知正三棱台 $ABC-A_1B_1C_1$ 的体积为 $\frac{52}{3}$，$AB=6$，$A_1B_1=2$，则 A_1A 与平面 ABC 所成角的正切值为

A：$\frac{1}{2}$　　B：1　　C：2　　D：3

答案：B.

解析：由题意知，$S_{\triangle A_1B_1C_1}=\sqrt{3}$，$S_{\triangle ABC}=9\sqrt{3}$. 易得 $A_1O_1=\frac{2\sqrt{3}}{3}$，$AO=2\sqrt{3}$，所以 $AM=\frac{4\sqrt{3}}{3}$.

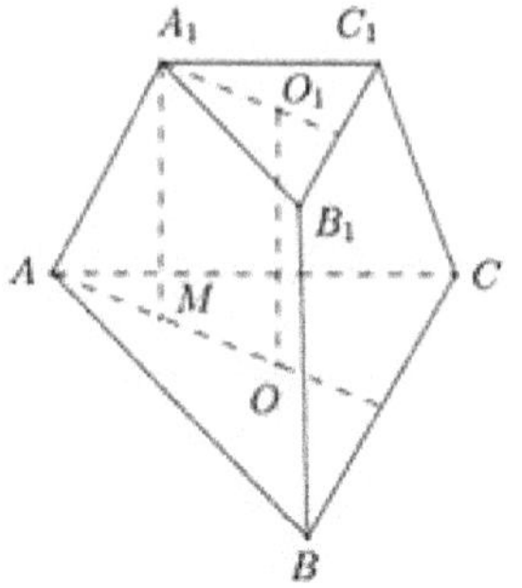

又 $V=\frac{1}{3}\cdot(\sqrt{3}+9\sqrt{3}+\sqrt{\sqrt{3}\cdot 9\sqrt{3}})\cdot OO_1=\frac{52}{3}$，所以 $A_1M=OO_1=\frac{4\sqrt{3}}{3}$.

所以，$\tan\angle A_1AM=1$. 故选 B.

Given a regular triangular prism ABC-$A_1B_1C_1$ with a volume of 32√3 cubic units. The lengths of the sides of the bases are AB = 6 and A_1B_1 = 2 units. Find the tangent of the angle between line segment A_1A and plane ABC.

A：$\frac{1}{2}$　　B：1　　C：2　　D：3

Answer: B.

Explanation:

From the problem we know $S_{\triangle A_1B_1C_1}=\sqrt{3}$, $S_{\triangle ABC}=9\sqrt{3}$. **we get** $A_1O_1=\frac{2\sqrt{3}}{3}$, $AO=2\sqrt{3}$, **so** $AM=\frac{4\sqrt{3}}{3}$.

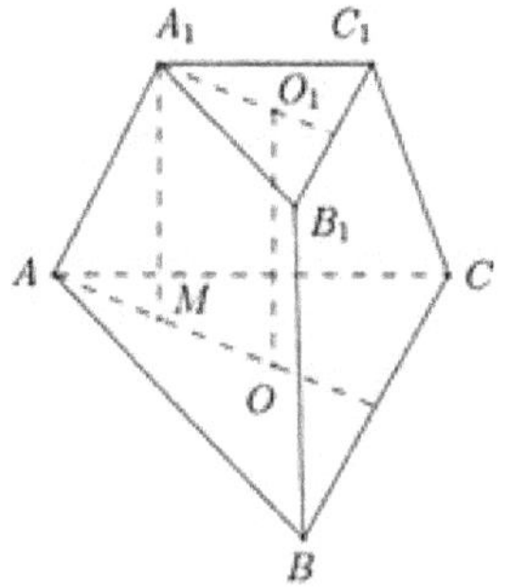

And $V=\frac{1}{3}\cdot(\sqrt{3}+9\sqrt{3}+\sqrt{\sqrt{3}\cdot 9\sqrt{3}})\cdot OO_1=\frac{52}{3}$，所以 $A_1M=OO_1=\frac{4\sqrt{3}}{3}$. **so** $\tan\angle A_1AM=1$.

Therefore, the correct answer is B.

8. 设函数 $f(x)=(x+a)\ln(x+b)$，若 $f(x)\geqslant 0$，则 a^2+b^2 的最小值为（ ）.

A: $\frac{1}{8}$　　B: $\frac{1}{4}$　　C: $\frac{1}{2}$　　D: 1

答案：C.

解析：$f(x)=(x+a)\ln(x+b)\ (x>-b)$. 令 $g(x)=x+a, h(x)=\ln(x+b)$，则 $f(x)=g(x)\cdot h(x)\geqslant 0$.

又 $g(x)$ 单调递增，$h(x)$ 单调递增，所以只需 $[-a,+\infty)$ 和 $[1-b,+\infty)$ 满足 $-a=1-b$，则

$$a^2+b^2=2b^2-2b+1,$$

其最小值为 $\frac{1}{2}$. 故选 C.

Given the function f(x) = (x+a)ln(x+b), find the minimum value of a^2+b^2 such that f(x) ≥ 0 for all x.

A: $\frac{1}{8}$　　B: $\frac{1}{4}$　　C: $\frac{1}{2}$　　D: 1

Answer: C.

Solution:

$f(x)=(x+a)\ln(x+b)\ (x>-b)$. **let** $g(x)=x+a, h(x)=\ln(x+b)$, 则 $f(x)=g(x)\cdot h(x)\geqslant 0$.

- **g(x) is monotonically increasing, h(x) is monotonically increasing, so we only need to consider the intervals [-a, +∞) and [1-b, +∞).**
- **Satisfying -a ≤ 1-b, then...** $a^2+b^2=2b^2-2b+1,$ **the minimum value is ½. Therefore, the answer is C.**

二、多项选择题：本题共 3 小题，每小题 6 分，共 18 分. 每小题给出的选项中，有多项符合题目要求. 全部选对的得 6 分，部分选对的得 3 分，选错或不选得 0 分.

II. Multiple Choice Questions: This section contains 3 questions, each worth 6 points, for a total of 18 points. For each question, there are multiple correct answers. Selecting all correct answers earns 6 points, partially correct answers earn 3 points, and incorrect or no answers earn 0 points.

9. 对于函数 $f(x)=\sin 2x$ 和 $g(x)=\sin(2x-\frac{\pi}{4})$，下列正确的有（ ）.

A: $f(x)$ 与 $g(x)$ 有相同的零点　　B: $f(x)$ 与 $g(x)$ 有相同的最大值

C: $f(x)$ 与 $g(x)$ 有相同的最小正周期　　D: $f(x)$ 与 $g(x)$ 的图像有相同的对称轴

答案：BC.

解析：分析如下：

	$f(x)=\sin 2x$	$g(x)=\sin(2x-\frac{\pi}{4})$
零点	$2x=k\pi, k\in\mathbf{Z}\Rightarrow(\frac{k\pi}{2},0), k\in\mathbf{Z}$	$2x-\frac{\pi}{4}=k\pi, k\in\mathbf{Z}\Rightarrow(\frac{\pi}{8}+\frac{k\pi}{2},0), k\in\mathbf{Z}$
对称轴	$2x=\frac{\pi}{2}+k\pi, k\in\mathbf{Z}\Rightarrow x=\frac{\pi}{4}+\frac{k\pi}{2}, k\in\mathbf{Z}$	$2x-\frac{\pi}{4}=\frac{\pi}{2}+k\pi, k\in\mathbf{Z}\Rightarrow x=\frac{3\pi}{8}+\frac{k\pi}{2}, k\in\mathbf{Z}$
最小正周期	$\frac{2\pi}{2}=\pi$	$\frac{2\pi}{2}=\pi$
最大值	1	1

Question 9:

For the functions f(x) = sin(2x) and g(x) = sin(2x - π/4), which of the following are correct?

- **A: f(x) and g(x) have the same zero points.**
- **B: f(x) and g(x) have the same maximum value.**
- **C: f(x) and g(x) have the same minimum positive period.**
- **D: The graphs of f(x) and g(x) have the same axis of symmetry.**

Answer: BC

Analysis:

	f(x) = sin(2x)	**g(x) = sin(2x - π/2)**
Zeros:	$2x=k\pi, k\in\mathbf{Z}\Rightarrow(\frac{k\pi}{2},0), k\in\mathbf{Z}$	$2x-\frac{\pi}{4}=k\pi, k\in\mathbf{Z}\Rightarrow(\frac{\pi}{8}+\frac{k\pi}{2},0), k\in\mathbf{Z}$

Axis of Symemtry (AOS)	$2x=\frac{\pi}{2}+k\pi,k\in\mathbf{Z}\Rightarrow x=\frac{\pi}{4}+\frac{k\pi}{2},k\in\mathbf{Z}$	$2x-\frac{\pi}{4}=\frac{\pi}{2}+k\pi,k\in\mathbf{Z}\Rightarrow x=\frac{3\pi}{8}+\frac{k\pi}{2},k\in\mathbf{Z}$
Same minimum positive period (SMPP)	$\frac{2\pi}{2}=\pi$	$\frac{2\pi}{2}=\pi$
Maximum value: 1		

10. 抛物线 $C:y^2=4x$ 的准线为 l, P 为 C 上动点, 过 P 作 $\odot A:x^2+(y-4)^2=1$ 的一条切线, Q 为切点, 过点 P 作 l 的垂线, 垂足为 B. 则(　).

A: l 与 $\odot A$ 相切

B: 当 P,A,B 三点共线时, $|PQ|=\sqrt{15}$

C: 当 $|PB|=2$ 时, $PA\perp AB$

D: 满足 $|PA|=|PB|$ 的点 A 有且仅有 2 个

答案: ABD.

解析: A: $y^2=4x,p=2,l:x=-1$.

又 $\odot A$ 半径为 1, 圆心为 $A(0,4)$, 所以 $d_{A-l}=1=r$, 所以 $\odot A$ 与 l 相切, A 正确.

B: P,A,B 三点共线时, $y_P=y_A=4$.

代入 $y^2=4x$ 中, $x_P=4$, 所以 $PA=4$, 所以 $PQ=\sqrt{PA^2-r^2}=\sqrt{15}$, B 正确.

C: $PB=2$ 时, $x_P=1,y_P=2$. 此时, $B(-1,2),P(1,2),A(0,4)$, $AP^2=AB^2=5,BP^2=4$.

因为 $AP^2+AB^2\neq BP^2$, 所以 PA 与 AB 不垂直, C 错误.

D: 因为 $PB=PF$, 所以 $PA=PB$ 时, $PA=PF$. 所以, 点 P 在 AF 中垂线上.

又 $A(0,4),F(1,0)$, 所以 AF 方程为 $x=4y-\frac{15}{2}$. 联立 $\begin{cases}x=4y-\frac{15}{2},\\ y^2=4x\end{cases}$ 得 $y^2-16y+30=0,\Delta>0$.

所以 AF 与抛物线 C 有两个交点. 故点 P 有且仅有两个, D 正确.

Question 10:

Parabola C: $y^2=4x$ has a directrix l. P is a moving point on C. A tangent line is drawn from P to circle A: $x^2+(y-4)^2=1$, intersecting the circle at point Q. The perpendicular line from point P intersects line l at point B. Then:

- **A: Circle A and line l are tangent to each other.**
- **B: When points P, A, and B are collinear, $|PQ|=\sqrt{15}$.**
- **C: When |PB| = 2, PA is perpendicular to AB.**
- **D: There are exactly two points A that satisfy |PA| = |PB|.**

Answer: ABD

Analysis:

A: $y^2=4x,p=2,l:x=-1$. **The circle A has center (0, 4) and radius 1.**
Therefore, the circle A and line l are tangent to each other. This statement A is correct.

B: When points P, A, and B are collinear, $y_P=y_A=4$. Substituting $y^2=4x$ xp=4, so PA=4, so $PQ=\sqrt{PA^2-r^2}=\sqrt{15}$. Therefore, B is correct.

When |PB| = 2, $x_P=1,y_P=2$. then $B(-1,2),P(1,2),A(0,4)$, $AP^2=AB^2=5,BP^2=4$.

As $AP^2+AB^2\neq BP^2$.. so PA and AB are not perpendicular. Therefore, C is not correct.

D: Because $PB=PF$, so shen $PA=PB$, $PA=PF$. Therefore, point P is on the Perpendicular median of AF.and $A(0,4),F(1,0)$, so function AF is $x=4y-\frac{15}{2}$. so $\begin{cases}x=4y-\frac{15}{2},\\ y^2=4x\end{cases}$

We get: $y^2-16y+30=0,\Delta>0$. Therefore, AF intersects with parabola C at two points, so there are exaxtly two points for P. D is correct.

11. 设函数 $f(x)=2x^3-3ax^2+1$，则（ ）.

A：当 $a>1$ 时，$f(x)$ 有三个零点

B：当 $a<0$ 时，$x=0$ 是 $f(x)$ 的极大值点

C：存在 a, b，使得 $x=b$ 为曲线 $y=f(x)$ 的对称轴

D：存在 a，使得点 $(1, f(1))$ 为曲线 $y=f(x)$ 的对称中心

答案：AD.

解析：$f(x)=2x^3-3ax^2+1$，$f'(x)=6x^2-6ax=6x(x-a)$. 令 $f'(x)=0$，$x_1=0$，$x_2=a$.

A: $a>1$ 时，$f(x)$ 在 $(-\infty,0)\nearrow(0,a)\searrow(a,+\infty)\nearrow$.

又 $x\to-\infty$ 时，$f(x)\to-\infty$，$x\to+\infty$ 时，$f(x)\to+\infty$，$f(0)=1>0$，$f(1)=3-3a<0$，所以 $f(a)<0$.

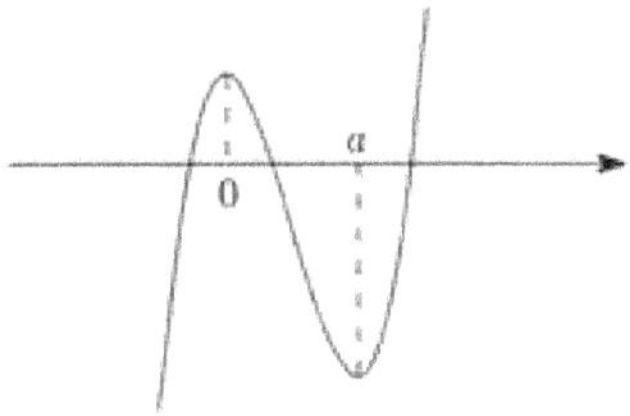

$f(x)$ 大致图像如图所示，所以有三个零点，A 正确.

B: $a<0$ 时，$f(x)$ 在 $(-\infty,a)\nearrow(a,0)\searrow(0,+\infty)\nearrow$，$x=0$ 为极小值点，B 错误.

C: 三次函数无对称轴，C 错误.

D: 令 $f(0)+f(2)=2f(1)$，即 $1+(2\times2^3-3a\times2^2+1)=2(3-3a)$，所以 $a=2$，

代入得 $f(x)=2x^3-6x^2+1$，满足 $f(x)+f(2-x)=2f(1)$，D 正确.

Problem 11

Let function $f(x)=2x^3-3ax^2+1$, then determine which of the following statements are true:

- **A: When a > 1, f(x) has three zeros.**
- **B: When a < 0, x = 0 is a local maximum point of f(x).**
- **C: There exist values of a and b such that x = b is the axis of symmetry of the curve y = f(x).**
- **D: There exists a value of a such that the point (1, f(1)) is the center of symmetry of the curve y = f(x).**

Answer: AD

Analysis:

$f(x)=2x^3-3ax^2+1$, $f'(x)=6x^2-6ax=6x(x-a)$.let $f'(x)=0$, $x_1=0$, $x_2=a$.

When a>1,f(x) at $(-\infty,0)\nearrow(0,a)\searrow(a,+\infty)\nearrow$. and when $x\to-\infty$, $f(x)\to-\infty$, when $x\to+\infty$ $f(x)\to+\infty$, $f(0)=1>0$, $f(1)=3-3a<0$, Therefore, f(a)<0. Therefore, the graph intersects the x-axis at three points, so there are exactly three zeros, and A is correct.

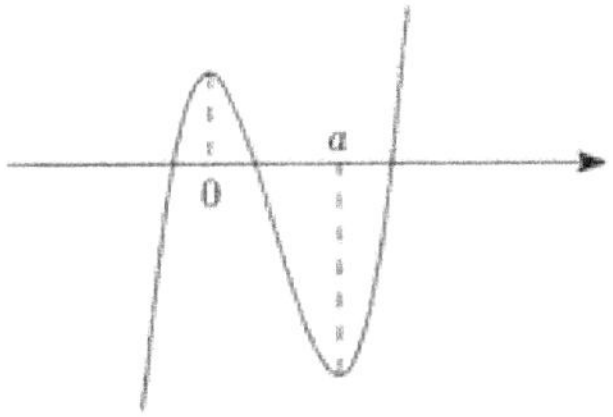

B, when a<0, f(x) at $(-\infty,a)\nearrow(a,0)\searrow(0,+\infty)\nearrow$, $x=0$ is minimum value, B is not correct.
C, A cubic function has no axis of symmetry. C is incorrect.
D: Let f(0) + f(2) = 2f(1), that is, 1 + (2 * 2^3 - 3a * 2^2 + 1) = 2(3 - 3 * a), so a = 2. Substituting, we get f(x) = 2x^3 - 6x^2 + 1, satisfying f(x) + f(2 - x) = 2f(1). Therefore, D is correct.

三、填空题：本题共 3 小题，每小题 5 分，共 15 分.

Section 3: Fill in the blanks. There are 3 questions in this section, 5 points for each question, and a total of 15 points.

12. 记 S_n 为等差数列 $\{a_n\}$ 的前 n 项和，若 $a_3+a_4=7, 3a_2+a_5=5$，则 $S_{10}=$ ________.

答案：95.

解析：$a_3+a_4=7, 3a_2+a_5=2a_2+a_3+a_4=5$，所以 $2a_2=-2, a_2=-1$.

又 $a_3+a_4=2a_2+3d=7$，所以 $d=3$，所以 $a_1=a_2-d=-4$. 故

$$S_{10}=10a_1+\frac{10\times 9}{2}\cdot d=10\times(-4)+45\times 3=95.$$

Question 12:
Let S_n be the sum of the first n terms of the arithmetic sequence $\{a_n\}$. If $a_3+a_4=7, 3a_2+a_5=5$, **find S_{10}.**
Answer: 95.
Solution:
$a_3+a_4=7, 3a_2+a_5=2a_2+a_3+a_4=5$, **so** $2a_2=-2, a_2=-1$. **also** $a_3+a_4=2a_2+3d=7$. **therefore** $2a_2=-2, a_2=-1$. **also** $a_3+a_4=2a_2+3d=7$, **thus d=3, so** $a_1=a_2-d=-4$. **Therefore:** $S_{10}=10a_1+\frac{10\times 9}{2}\cdot d=10\times(-4)+45\times 3=95$.

13. 已知 α 为第一象限角，β 为第三象限角，$\tan\alpha+\tan\beta=4, \tan\alpha\tan\beta=\sqrt{2}+1$，则 $\sin(\alpha+\beta)=$ ________.

答案：$-\frac{2\sqrt{2}}{3}$.

解析：因为 $\tan\alpha+\tan\beta=4, \tan\alpha\cdot\tan\beta=\sqrt{2}+1$，所以

$$\tan(\alpha+\beta)=\frac{\tan\alpha+\tan\beta}{1-\tan\alpha\cdot\tan\beta}=\frac{4}{1-(\sqrt{2}+1)}=-2\sqrt{2}.$$

又 α、β 分别为第一、三象限角，所以 $\begin{cases}2k\pi<\alpha<\frac{\pi}{2}+2k\pi, k\in\mathbf{Z},\\ \pi+2k\pi<\beta<\frac{3}{2}\pi+2k\pi, k\in\mathbf{Z},\end{cases}$ 所以

$$\pi+2k\pi<\alpha+\beta<2\pi+2k\pi, k\in\mathbf{Z}.$$

所以，$\alpha+\beta$ 为第三、四象限角.

又 $\tan(\alpha+\beta)=-2\sqrt{2}<0$，所以 $\alpha+\beta$ 为第四象限角，所以 $\sin(\alpha+\beta)<0, \cos(\alpha+\beta)>0$.

又 $\begin{cases}\tan(\alpha+\beta)=\frac{\sin(\alpha+\beta)}{\cos(\alpha+\beta)}=-2\sqrt{2},\\ \sin^2(\alpha+\beta)+\cos^2(\alpha+\beta)=1,\end{cases}$ 所以 $\sin(\alpha+\beta)=-\frac{2\sqrt{2}}{3}$.

Question 13:
Given that α is in the first quadrant and β is in the third quadrant, and tanα + tanβ = 4, tanα * tanβ = √2 + 1, find sin(α + β).
Answer: -2√2 / 3
Solution:
Because tanα + tanβ = 4 and tanα * tanβ = √2 + 1, we have:
tan(α + β) = (tanα + tanβ) / (1 - tanα * tanβ) = 4 / (1 - (√2 + 1)) = -2√2.
Since α is in the first quadrant and β is in the third quadrant,
Therefore, $\begin{cases}2k\pi<\alpha<\frac{\pi}{2}+2k\pi, k\in\mathbf{Z},\\ \pi+2k\pi<\beta<\frac{3}{2}\pi+2k\pi, k\in\mathbf{Z},\end{cases}$ **therefore,** $\pi+2k\pi<\alpha+\beta<2\pi+2k\pi, k\in\mathbf{Z}.$ **which means α + β is in the third or fourth quadrant. Given that tan(α + β) = -2√2 < 0, α + β must be in the fourth quadrant. So sin(α + β) < 0 and cos(α + β) > 0.**

We know $\begin{cases} \tan(\alpha+\beta) = \dfrac{\sin(\alpha+\beta)}{\cos(\alpha+\beta)} = -2\sqrt{2}, \\ \sin^2(\alpha+\beta) + \cos^2(\alpha+\beta) = 1, \end{cases}$ so we can find sin(α + β) = -2√2 / 3.

11	21	31	40
12	22	33	42
13	22	33	43
15	24	34	44

14. 在上图的 4×4 方格表中选 4 个方格，要求每行和每列均恰有一个方格被选中，则共有__________种选法，在所有符合上述要求的选法中，选中方格中的 4 个数之和的最大值是__________.

答案：24; 112.

解析：(1) 在四列中分别取一格，分别取第一、二、三、四行中的某一格，即相当于把取出的格子排序.

故 $A_4^4 = 24$ 种选法.

11	21	31	40
12	22	33	42
13	22	33	43
15	24	34	44

Question 14:

In the 4x4 grid above, select 4 squares with the requirement that there is exactly one selected square in each row and column. How many different ways are there to select? Among all the selections that meet the above requirements, what is the maximum sum of the 4 numbers in the selected squares?

Answer: 24; 112.

Explanation:

(1) Choose one square from each column, which is equivalent to sorting the selected squares by row (first, second, third, or fourth row). Therefore, there are 24 ways to select.

四、解答题：本题共 5 小题，共 77 分．解答应写出文字说明、证明过程或演算步骤．

Section 4: Problem Solving This section contains 5 questions, totaling 77 points. Answers must include explanations, proofs, or calculation steps.

15.(13 分)

记 $\triangle ABC$ 的内角 A,B,C 的对边分别为 a,b,c，已知 $\sin A+\sqrt{3}\cos A=2$.

(1) 求 A.

(2) 若 $a=2$, $\sqrt{2}b\sin C=c\sin 2B$, 求 $\triangle ABC$ 的周长.

解：$\because\ \sin A+\sqrt{3}\cos A=2$, $\therefore\ 2\sin(A+\frac{\pi}{3})=2$, $\sin(A+\frac{\pi}{3})=1$.

又 $A\in(0,\pi)$, $\therefore\ A+\frac{\pi}{3}=\frac{\pi}{2}$, $A=\frac{\pi}{6}$.

综上，角 A 为 $\frac{\pi}{6}$.

(2) 若 $a=2$, $\sqrt{2}b\sin C=c\sin 2B$, 求 $\triangle ABC$ 的周长.

解：$\because\ \sin A+\sqrt{3}\cos A=2$, $\therefore\ 2\sin(A+\frac{\pi}{3})=2$, $\sin(A+\frac{\pi}{3})=1$.

又 $A\in(0,\pi)$, $\therefore\ A+\frac{\pi}{3}=\frac{\pi}{2}$, $A=\frac{\pi}{6}$.

综上，角 A 为 $\frac{\pi}{6}$.

(2) $\because\ \sqrt{2}b\sin c=c\sin 2B$, $\therefore\ \sqrt{2}b\sin C=2c\sin B\cos B$, $\therefore\ \sqrt{2}bc=2bc\cos B$, $\cos B=\frac{\sqrt{2}}{2}$.

又 $B\in(0,\pi)$, $\therefore\ B=\frac{\pi}{4}$, $C=\pi-A-B=\frac{7}{12}\pi$. 在 $\triangle ABC$ 中，由正弦定理得

$$\frac{a}{\sin A}=\frac{b}{\sin B}=\frac{c}{\sin C}=\frac{2}{\frac{1}{2}}=4,$$

$\therefore\ b=4\sin B=2\sqrt{2}$, $c=4\sin C=4\sin\frac{7\pi}{12}=4\sin(\frac{\pi}{4}+\frac{\pi}{3})=\sqrt{6}+\sqrt{2}$. $\therefore\ a+b+c=2+3\sqrt{2}+\sqrt{6}$.

综上，$\triangle ABC$ 的周长为 $2+3\sqrt{2}+\sqrt{6}$.

Let A, B, and C be the interior angles of triangle ABC, with corresponding opposite sides a, b, and c respectively. Given sin A + √3 cos A = 2.

(1) Find angle A.

(2) If a = 2, √2b sin C = c sin 2B, find the perimeter of triangle ABC.

(1) $\because\ \sin A+\sqrt{3}\cos A=2$, $\therefore\ 2\sin(A+\frac{\pi}{3})=2$, $\sin(A+\frac{\pi}{3})=1$. **and** $A\in(0,\pi)$, $\therefore\ A+\frac{\pi}{3}=\frac{\pi}{2}$, $A=\frac{\pi}{6}$.

so angle A= $\frac{\pi}{6}$.

(2) $\because\ \sqrt{2}b\sin c=c\sin 2B$, $\therefore\ \sqrt{2}b\sin C=2c\sin B\cos B$, $\therefore\ \sqrt{2}bc=2bc\cos B$, $\cos B=\frac{\sqrt{2}}{2}$.

and $B\in(0,\pi)$, $\therefore\ B=\frac{\pi}{4}$, $C=\pi-A-B=\frac{7}{12}\pi$. **within** $\triangle ABC$ **by Law of Sines:**

$$\frac{a}{\sin A}=\frac{b}{\sin B}=\frac{c}{\sin C}=\frac{2}{\frac{1}{2}}=4,$$

$\therefore\ b=4\sin B=2\sqrt{2}$, $c=4\sin C=4\sin\frac{7\pi}{12}=4\sin(\frac{\pi}{4}+\frac{\pi}{3})=\sqrt{6}+\sqrt{2}$. $\therefore\ a+b+c=2+3\sqrt{2}+\sqrt{6}$.

Therefore, the perimeter of triangle ABC is $2+3\sqrt{2}+\sqrt{6}$.

16. (15 分)

已知函数 $f(x) = e^x - ax - a^3$.

(1) 当 $a = 1$ 时, 求曲线 $y = f(x)$ 在点 $(1, f(1))$ 处的切线方程.

(2) 若 $f(x)$ 有极小值, 且极小值小于 0, 求 a 的取值范围.

解: (1) 当 $a = 1$ 时, $f(x) = e^x - x - 1$, $f'(x) = e^x - 1$.

令 $x = 1$, 得 $f(1) = e - 2$, $f'(1) = e - 1$.

故 $f(x)$ 在 $(1, f(1))$ 处的切线方程为 $(e - 1)(x - 1) = y - (e - 2)$, 整理得 $(e - 1)x - y - 1 = 0$.

综上, 曲线 $y = f(x)$ 在 $(1, f(1))$ 处的切线方程为 $(e - 1)x - y - 1 = 0$.

(2) 因为 $f(x) = e^x - ax - a^3$, 所以 $f(x)$ 定义域为 $\mathbf{R}$, 且 $f'(x) = e^x - a$, $f'(x)$ 在 $\mathbf{R}$ 上单调递增.

当 $a \leqslant 0$ 时, $\forall x \in \mathbf{R}$, $f'(x) > 0$ 恒成立, $f(x)$ 无极小值.

当 $a > 0$ 时, 令 $f'(x) = 0$ 得 $x = \ln a$.

所以, 当 $x \in (-\infty, \ln a)$ 时, $f'(x) < 0$, $f(x)$ 单调递减; 当 $x \in (\ln a, +\infty)$ 时, $f'(x) > 0$, $f(x)$ 单调递增.

即 $f(x)$ 在 $x = \ln a$ 处取极小值, 极小值 $f(\ln a) = a - a\ln a - a^3$.

又 $f(x)$ 的极小值小于 0, 所以 $a - a\ln a - a^3 < 0$, 即 $a^2 + \ln a - 1 > 0$.

令 $g(a) = a^2 + \ln a - 1$, 则 $g'(a) = 2a + \frac{1}{a} > 0$, $g(a)$ 单调递增.

又 $g(1) = 1^2 + \ln 1 - 1 = 0$, 所以 $a^2 + \ln a - 1 > 0$ 的解集为 $a \in (1, +\infty)$.

综上, a 的取值范围为 $(1, +\infty)$.

Question 16 (15 points)
Given the function $f(x) = e^x - ax - a^3$.
(1) When a = 1, find the equation of the tangent line to the curve y = f(x) at point (1, f(1)).
(2) If f(x) has a minimum value and the minimum value is less than 0, find the range of values for a.
Solution:
(1) When a = 1, $f(x) = e^x - x - 1$, $f'(x) = e^x - 1$.
Let x = 1, we get $f(1) = e - 2$, $f'(1) = e - 1$.
Therefore, the equation of the tangent line to f(x) at (1, f(1)) is $(e - 1)(x - 1) = y - (e - 2)$, which simplifies to (e-1)x - y - 1 = 0.
In summary, the equation of the tangent line to y = f(x) at (1, f(1)) is (e-1)x - y - 1 = 0.
(2) Because $f(x) = e^x - ax - a^3$, so the domain of f(x) is R, and f'(x) = e^x - a, f'(x) is increasing on R.
When a ≤ 0, for all x ∈ R, f'(x) > 0 always holds, and f(x) has no minimum value.
When a > 0, let f'(x) = 0, we get x = ln a.
Therefore, when x ∈ (-∞, ln a), f'(x) < 0, f(x) is decreasing; when x ∈ (ln a, +∞), f'(x) > 0, f(x) is increasing.
That is, f(x) has a minimum value at x = ln a, and the minimum value is f(ln a) = a – a*ln a - a^3.
Since the minimum value of f(x) is less than 0, we have a - aln a - a^3 < 0, which means a^2 + ln a - 1 > 0.
Let g(a) = a^2 + ln a - 1, then g'(a) = 2a + 1/a > 0, g(a) is increasing.
Also, g(1) = 1^2 + ln 1 - 1 = 0, so the solution to a^2 + ln a - 1 > 0 is a ∈ (1, +∞).
In summary, the range of values for a is (1, +∞).

17. (15 分)

如图, 平面四边形 $ABCD$ 中, $AB = 8$, $CD = 3$, $AD = 5\sqrt{3}$, $\angle ADC = 90°$, $\angle BAD = 30°$, 点 E, F 满足 $\overrightarrow{AE} = \frac{2}{5}\overrightarrow{AD}$, $\overrightarrow{AF} = \frac{1}{2}\overrightarrow{AB}$. 将 $\triangle AEF$ 沿 EF 翻折至 $\triangle PEF$, 使得 $PC = 4\sqrt{3}$.

(1) 证明: $EF \perp PD$.

(2) 求面 PCD 与面 PBF 所成的二面角的正弦值.

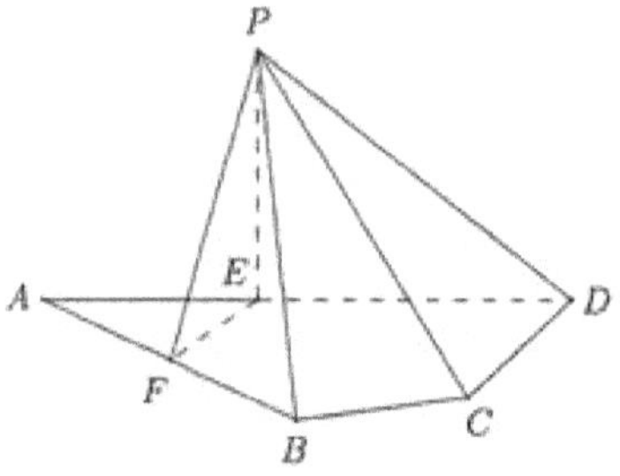

题图

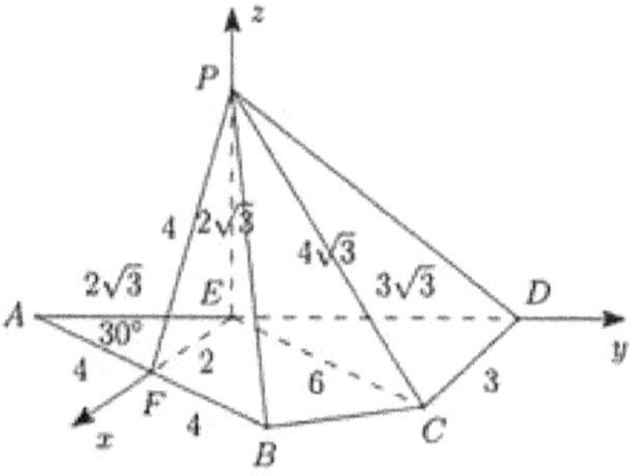

解析图

解: (1) 连接 EC, 在 $\triangle AEF$ 中, 由余弦定理知 $EF = 2$, 则 $EF \perp AE$.

$\therefore EF \perp PE, EF \perp ED$, 则 $EF \perp$ 平面 PED, $\therefore EF \perp PD$.

(2) $\triangle CDE$ 中, $CE = \sqrt{DE^2 + CD^2} = \sqrt{27+9} = 6$; $\triangle PCE$ 中, $PE^2 + CE^2 = PC^2$, $\therefore PE \perp EC$.

易知 EP、EF、ED 两两垂直. 以 EF、ED、EP 所在直线分别为 x 轴、y 轴、z 轴建立空间直角坐标系.

则 $P(0,0,2\sqrt{3})$, $F(2,0,0)$, $B(4,2\sqrt{3},0)$, $C(3,3\sqrt{3},0)$, $D(0,3\sqrt{3},0)$.

$\overrightarrow{PB} = (2,0,-2\sqrt{3})$, $\overrightarrow{FB} = (2,2\sqrt{3},0)$, 可求得平面 PBF 的一个法向量 $\overrightarrow{n_1} = (\sqrt{3},-1,1)$.

$\overrightarrow{PD} = (0,3\sqrt{3},-2\sqrt{3})$, $\overrightarrow{CD} = (-3,0,0)$, 可求得平面 PCD 的一个法向量 $\overrightarrow{n_2} = (0,2,3)$.

所以, $\cos\theta = \left|\frac{\overrightarrow{n_1} \cdot \overrightarrow{n_2}}{|\overrightarrow{n_1}||\overrightarrow{n_2}|}\right| = \frac{1}{\sqrt{5}\cdot\sqrt{13}} = \frac{1}{\sqrt{65}}$, $\sin\theta = \frac{8\sqrt{65}}{65}$.

Question 17 (15 points)

As shown in the figure, in quadrilateral ABCD, AB = 8, CD = 3, $AD = 5\sqrt{3}$, **∠ADC = 90°, ∠BAD = 30°. Points E and F satisfy** $\overrightarrow{AE} = \frac{2}{5}\overrightarrow{AD}, \overrightarrow{AF} = \frac{1}{2}\overrightarrow{AB}$. **Triangle AEF is folded along EF to triangle PEF, such that PC = 4√3.**

Prove: EF is perpendicular to PD.

Find: The sine of the dihedral angle between planes PCD and PBF.

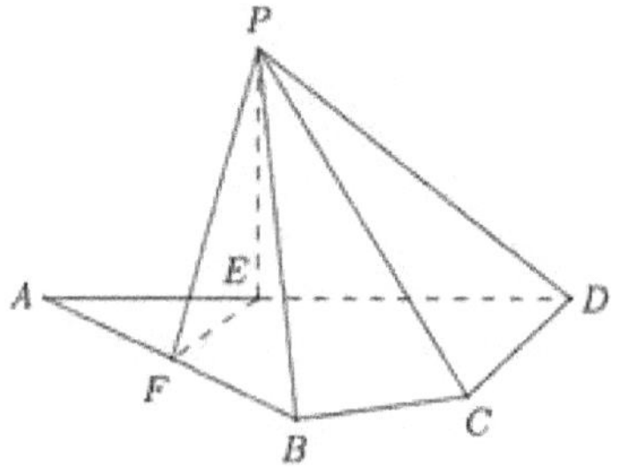

题图

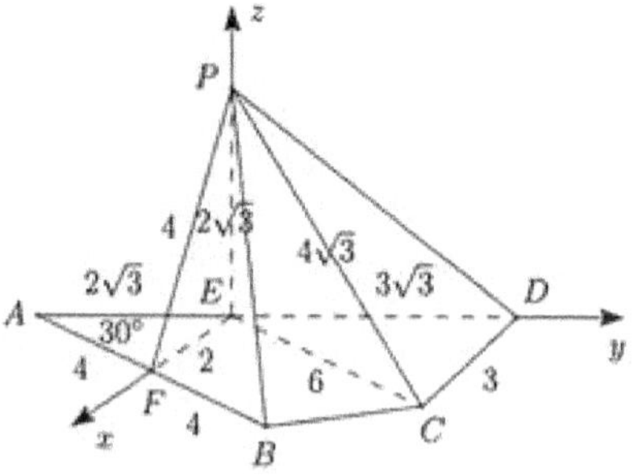

解析图

A diagram: This is often labeled "题图" (title image) or "解析图" (analysis diagram). It visualizes the geometric problem.

解：(1) 连接 EC，在 $\triangle AEF$ 中，由余弦定理知 $EF=2$，则 $EF\perp AE$.

$\therefore EF\perp PE, EF\perp ED$，则 $EF\perp$ 平面 PED，$\therefore EF\perp PD$.

(2) $\triangle CDE$ 中，$CE=\sqrt{DE^2+CD^2}=\sqrt{27+9}=6$；$\triangle PCE$ 中，$PE^2+CE^2=PC^2$，$\therefore PE\perp EC$.

易知 EP、EF、ED 两两垂直. 以 EF、ED、EP 所在直线分别为 x 轴、y 轴、z 轴建立空间直角坐标系.

则 $P(0,0,2\sqrt{3})$, $F(2,0,0)$, $B(4,2\sqrt{3},0)$, $C(3,3\sqrt{3},0)$, $D(0,3\sqrt{3},0)$.

$\overrightarrow{PB}=(2,0,-2\sqrt{3}), \overrightarrow{FB}=(2,2\sqrt{3},0)$，可求得平面 PBF 的一个法向量 $\overrightarrow{n_1}=(\sqrt{3},-1,1)$.

$\overrightarrow{PD}=(0,3\sqrt{3},-2\sqrt{3}), \overrightarrow{CD}=(-3,0,0)$，可求得平面 PCD 的一个法向量 $\overrightarrow{n_2}=(0,2,3)$.

所以，$\cos\theta=|\frac{\overrightarrow{n_1}\cdot\overrightarrow{n_2}}{|\overrightarrow{n_1}||\overrightarrow{n_2}|}|=\frac{1}{\sqrt{5}\cdot\sqrt{13}}=\frac{1}{\sqrt{65}}, \sin\theta=\frac{8\sqrt{65}}{65}$.

In $\triangle AEF$ connect EC, by Law of Cosines, we know $EF=2$, so $EF\perp AE$. $\therefore EF\perp PE, EF\perp ED$, then EF is perpendicular to plane PED, $\therefore EF\perp PD$.

In $\triangle CDE$, $CE=\sqrt{DE^2+CD^2}=\sqrt{27+9}=6$; $\triangle PCE$ 中, $PE^2+CE^2=PC^2$, $\therefore PE\perp EC$

It is easy to see that EP, EF, and ED are mutually perpendicular. Establish a spatial rectangular coordinate system with the lines containing EF, ED, and EP as the x-axis, y-axis, and z-axis respectively. We find: $P(0,0,2\sqrt{3})$, $F(2,0,0)$, $B(4,2\sqrt{3},0)$, $C(3,3\sqrt{3},0)$, $D(0,3\sqrt{3},0)$.

$\overrightarrow{PB}=(2,0,-2\sqrt{3}), \overrightarrow{FB}=(2,2\sqrt{3},0)$, a normal vector of plane PBF can be found as: $\overrightarrow{n_1}=(\sqrt{3},-1,1)$. $\overrightarrow{PD}=(0,3\sqrt{3},-2\sqrt{3}), \overrightarrow{CD}=(-3,0,0)$, a normal vector of plane PCD can be found as: $\overrightarrow{n_2}=(0,2,3)$, therefore,

$$\cos\theta=|\frac{\overrightarrow{n_1}\cdot\overrightarrow{n_2}}{|\overrightarrow{n_1}||\overrightarrow{n_2}|}|=\frac{1}{\sqrt{5}\cdot\sqrt{13}}=\frac{1}{\sqrt{65}}, \sin\theta=\frac{8\sqrt{65}}{65}.$$

18. (17 分)

某投篮比赛分为两个阶段，每个参赛队由两名队员组成，比赛具体规则如下：第一阶段由参赛队中一名队员投篮 3 次，若 3 次都未投中，则该队被淘汰，比赛成绩为 0 分；若至少投中一次，则该队进入第二阶段，由该队的另一名队员投篮 3 次，每次投中得 5 分，未投中得 0 分，该队的比赛成绩为第二阶段的得分总和.

某参赛队由甲、乙两名队员组成，设甲每次投中的概率为 p，乙每次投中的概率为 q，各次投中与否相互独立.

(1) 若 $p=0.4, q=0.5$，甲参加第一阶段比赛，求甲、乙所在队的比赛成绩不少于 5 分的概率.

(2) 假设 $0<p<q$.

(i) 为使得甲、乙所在队的比赛成绩为 15 分的概率最大，应该由谁参加第一阶段的比赛?

(ii) 为使得甲、乙所在队的比赛成绩的数学期望最大，应该由谁参加第一阶段的比赛?

A basketball shooting competition consists of two stages. Each team has two members. The specific rules of the competition are as follows:

- **Stage 1: One member of the team shoots three times. If all three shots miss, the team is eliminated and scores 0 points. If at least one shot is made, the team advances to the second stage.**
- **Stage 2: The other member of the team shoots three times. Each successful shot scores 5 points, and each missed shot scores 0 points. The team's final score is the total score from the second stage.**

A certain team consists of two members, A and B. Let p be the probability of A making a shot, and q be the probability of B making a shot. Assume that the outcomes of each shot are independent.

(1) If p = 0.4 and q = 0.5, and A participates in the first stage, find the probability that the team's final score is at least 5 points.

(2) Assume 0 < p < q.

(i) To maximize the probability of the team scoring 15 points, who should participate in the first stage?

(ii) To maximize the expected value of the team's score, who should participate in the first stage?

解：(1) 设甲、乙所在队的比赛成绩不少于 5 为事件 A, 则甲在第一阶段至少投中一次, 乙在第二阶段至少投中一次. $P(A) = (1 - 0.6^3)(1 - 0.5^3) = 0.686$.

综上, 甲、乙所在队的比赛成绩不少于 5 分的概率为 0.686.

(2) (i) 设第一阶段由甲比赛, 且比赛成绩为 15 分为事件 B, 第一阶段由乙比赛, 且比赛成绩为 15 分为事件 C.

$$P(B) = [1 - (1 - p)^3]q^3, \quad P(C) = [1 - (1 - q)^3]p^3,$$

$$P(B) - P(C) = [1 - (1 - p)^3]q^3 - [1 - (1 - q)^3]p^3 = 3pq(p + q - pq)(q - p)$$

$$= 3pq[1 - (1 - p)(1 - q)](q - p) > 0.$$

综上, 由甲参加第一阶段的比赛比赛成绩为 15 分的概率最大.

(ii) 设第一阶段由甲参赛, 所在队最终成绩为 X, 第一阶段由乙参赛, 所在队最终成绩为 Y.

则 $X = 0, 5, 10, 15; Y = 0, 5, 10, 15$.

$$P(X = 0) = (1 - p)^3 + [1 - (1 - p)^3](1 - q)^3$$

$$P(X = 5) = [1 - (1 - p)^3] \times 3q(1 - q)^2$$

$$P(X = 10) = [1 - (1 - p)^3] \times 3q^2(1 - q)$$

$$P(X = 15) = [1 - (1 - p)^3]q^3$$

$$E(X) = 0 \times P(X = 0) + 5 \times P(X = 5) + 10 \times P(X = 10) + 15 \times P(X = 15)$$

$$= 15[1 - (1 - p)^3]q(1 - q)^2 + 30[1 - (1 - p)^3]q^2(1 - q) + 15[1 - (1 - p)^3]q^3$$

$$= 15q[1 - (1 - p)^3][(1 - q)^2 + 2q(1 - q) + q^2]$$

$$= 15q[1 - (1 - p)^3].$$

同理, $E(Y) = 15p[1 - (1 - q)^3]$.

所以, $E(X) - E(Y) = 15q[1 - (1 - p)^3] - 15p[1 - (1 - q)^3] = 15pq(p + q - 3)(p - q) > 0$.

故为使甲乙所在队成绩数学期望最大, 应由甲参加一阶段比赛.

Solution:

(1) Let event A be the event that the team's score is at least 5 points. Then, A occurs when player A makes at least one shot in the first stage and player B makes at least one shot in the second stage. P(A) = (1 - 0.6^3)(1 - 0.5^3) = 0.686.

Therefore, the probability that the team's score is at least 5 points is 0.686.

(2) (i) Let event B be the event that the team scores 15 points when player A plays in the first stage, and event C be the event that the team scores 15 points when player B plays in the first stage.

$$P(B) = [1 - (1 - p)^3]q^3, \quad P(C) = [1 - (1 - q)^3]p^3,$$

$$P(B) - P(C) = [1 - (1 - p)^3]q^3 - [1 - (1 - q)^3]p^3 = 3pq(p + q - pq)(q - p)$$

$$= 3pq[1 - (1 - p)(1 - q)](q - p) > 0.$$

In summary, when player A participates in the first stage, the probability of the team scoring 15 points is higher. Let X be the final score of the team when player A participates in the first stage, and Y be the final score of the team when player B participates in the first stage.

Then $X = 0, 5, 10, 15; Y = 0, 5, 10, 15.$

$$P(X=0)=(1-p)^3+[1-(1-p)^3](1-q)^3$$

$$P(X=5)=[1-(1-p)^3]\times 3q(1-q)^2$$

$$P(X=10)=[1-(1-p)^3]\times 3q^2(1-q)$$

$$P(X=15)=[1-(1-p)^3]q^3$$

$$\begin{aligned}E(X)&=0\times P(X=0)+5\times P(X=5)+10\times P(X=10)+15\times P(X=15)\\&=15[1-(1-p)^3]q(1-q)^2+30[1-(1-p)^3]q^2(1-q)+15[1-(1-p)^3]q^3\\&=15q[1-(1-p)^3][(1-q)^2+2q(1-q)+q^2]\\&=15q[1-(1-p)^3].\end{aligned}$$

Similarly, $E(Y)=15p[1-(1-q)^3]$.

Therefore: $E(X)-E(Y)=15q[1-(1-p)^3]-15p[1-(1-q)^3]=15pq(p+q-3)(p-q)>0.$

Therefore, to maximize the expected value of the team's score, player A should participate in the first stage.

19. (17 分)

已知双曲线 $C: x^2-y^2=m\ (m>0)$, 点 $P_1(5,4)$ 在 C 上, k 为常数, $0<k<1$. 按照如下方式依次构造点 $P_n\ (n=2,3,\cdots)$, 过点 P_{n-1} 作斜率为 k 的直线与 C 的左支交于点 Q_{n-1}, 令 P_n 为 Q_{n-1} 关于 y 轴的对称点, 记 P_n 的坐标为 (x_n, y_n).

(1) 若 $k=\frac{1}{2}$, 求 x_2, y_2.

(2) 证明: 数列 $\{x_n-y_n\}$ 是公比为 $\frac{1+k}{1-k}$ 的等比数列.

(3) 设 S_n 为 $\triangle P_nP_{n+1}P_{n+2}$ 的面积, 证明: 对任意的正整数 n, $S_n=S_{n+1}$.

解: (1) 因为 $P_1(5,4)$ 在 C 上, 所以 $m=5^2-4^2=9$. 故双曲线方程为 $C: \frac{x^2}{9}-\frac{y^2}{9}=1$.

由已知有 $l_{P_1Q_1}: y-4=\frac{1}{2}(x-5)$, 即 $y=\frac{1}{2}x+\frac{3}{2}$, 与 C 联立有 $y(y-4)=0$, 所以 $Q_1(-3,0)$, 则 $P_2(3,0)$.

所以, $x_2=3, y_2=0$.

(2) 点 $P_n(x_n,y_n)$, $P_{n+1}(x_{n+1},y_{n+1})$, $Q_n(-x_{n+1},y_{n+1})$ 满足:

$$\begin{cases}x_n^2-y_n^2=9,\\x_{n+1}^2-y_{n+1}^2=9,\end{cases}\qquad\begin{cases}(x_n-y_n)(x_n+y_n)=9,\\(x_{n+1}-y_{n+1})(x_{n+1}+y_{n+1})=9.\end{cases}$$

且 $y_{n+1}-y_n=k(-x_{n+1}-x_n)$, $k=\frac{y_n-y_{n+1}}{x_{n+1}+x_n}$.

所以,

$$\frac{1+k}{1-k}=\frac{1+\frac{y_n-y_{n+1}}{x_{n+1}+x_n}}{1-\frac{y_n-y_{n+1}}{x_{n+1}+x_n}}=\frac{(x_{n+1}-y_{n+1})+(x_n+y_n)}{(x_{n+1}+y_{n+1})+(x_n-y_n)}=\frac{(x_{n+1}-y_{n+1})+\frac{9}{x_n-y_n}}{(x_n-y_n)+\frac{9}{x_{n+1}-y_{n+1}}}$$

$$=\frac{\frac{1}{x_n-y_n}\cdot[(x_{n+1}-y_{n+1})(x_n-y_n)+9]}{\frac{1}{x_{n+1}-y_{n+1}}\cdot[(x_{n+1}-y_{n+1})(x_n-y_n)+9]}=\frac{x_{n+1}-y_{n+1}}{x_n-y_n}.$$

故 $\{x_n-y_n\}$ 为等比数列, 且公比为 $\frac{1+k}{1-k}$.

Given the hyperbola C: $x^2/m - y^2/m = 1$ (m > 0), point $P_1(5,4)$ **is on C, and k is a constant, 0 < k < 1. Points P_n (n = 2, 3, ...) are constructed as follows:**

- **Draw a line with slope k through point P_{n-1} that intersects the left branch of C at point Q_{n-1}.**
- **Let P_n be the symmetric point of Q_{n-1} with respect to the y-axis.**
- **Denote the coordinates of P_n as** (x_n, y_n).
  - **(1) If k = ½, find** x_2, y_2.

(2) Prove the sequence $\{x_n - y_n\}$ is a geometric sequence with common ratio $\frac{1+k}{1-k}$

(3) Let S_n be the area of triangle $P_nP_{n+1}P_{n+2}$. Prove: For any positive integer n, $S_n = S_{n+1}$

Solution: (1) as $P_1(5,4)$ is on C, so $m = 5^2 - 4^2 = 9$. so Hyperbola equation is $C: \frac{x^2}{9} - \frac{y^2}{9} = 1$.

We know: $l_{P_1Q_1}: y - 4 = \frac{1}{2}(x-5)$, which means $y = \frac{1}{2}x + \frac{3}{2}$, combine C we get $y(y-4) = 0$, so $Q_1(-3,0)$, then $P_2(3,0)$, therefore, $x_2 = 3, y_2 = 0$.

(3) Point $P_n(x_n, y_n), P_{n+1}(x_{n+1}, y_{n+1}), Q_n(-x_{n+1}, y_{n+1})$ satisfy:

$$\begin{cases} x_n^2 - y_n^2 = 9, \\ x_{n+1}^2 - y_{n+1}^2 = 9. \end{cases} \quad \begin{cases} (x_n - y_n)(x_n + y_n) = 9, \\ (x_{n+1} - y_{n+1})(x_{n+1} + y_{n+1}) = 9. \end{cases} \quad 且\ y_{n+1} - y_n = k(-x_{n+1} - x_n), k = \frac{y_n - y_{n+1}}{x_{n+1} + x_n}.$$

Therefore,

$$\frac{1+k}{1-k} = \frac{1 + \frac{y_n - y_{n+1}}{x_{n+1} + x_n}}{1 - \frac{y_n - y_{n+1}}{x_{n+1} + x_n}} = \frac{(x_{n+1} - y_{n+1}) + (x_n + y_n)}{(x_{n+1} + y_{n+1}) + (x_n - y_n)} = \frac{(x_{n+1} - y_{n+1}) + \frac{9}{x_n - y_n}}{(x_n - y_n) + \frac{9}{x_{n+1} - y_{n+1}}}$$

$$= \frac{\frac{1}{x_n - y_n} \cdot [(x_{n+1} - y_{n+1})(x_n - y_n) + 9]}{\frac{1}{x_{n+1} - y_{n+1}} \cdot [(x_{n+1} - y_{n+1})(x_n - y_n) + 9]} = \frac{x_{n+1} - y_{n+1}}{x_n - y_n}.$$

Therefore, the sequence $\{x_n - y_n\}$ is a geometric sequence with common ratio $\frac{1+k}{1-k}$

全国甲卷

绝密★启用前

2024 年普通高等学校招生全国统一考试

全国甲卷理科数学

使用范围：陕西、宁夏、青海、内蒙古、四川

注意事项：

1. 答题前，务必将自己的姓名、考籍号填写在答题卡规定的位置上.
2. 答选择题时，必须使用 2B 铅笔将答题卡上对应题目的答案标号涂黑. 如需改动，用橡皮擦擦干净后，再选涂其它答案标号.
3. 答非选择题时，必须使用 0.5 毫米黑色签字笔，将答案书写在答题卡规定的位置上.
4. 所有题目必须在答题卡上作答，在试题卷上答题无效.
5. 考试结束后，只将答题卡交回.

一、选择题：本题共 12 小题，每小题 5 分，共 60 分. 在每小题给出的四个选项中，只有一项是符合题目要求的.

National Volume ATop Secret★Do Not Open Until Authorized2024 National Unified Entrance Examination for Ordinary Colleges and UniversitiesNational Volume A Mathematics for ScienceScope of Use: Shaanxi, Ningxia, Qinghai, Inner Mongolia, Sichuan

In essence, this image is a notice for the 2024 National College Entrance Examination (Gaokao) for Mathematics (Science) for the National Volume A (全国甲卷), which is used in the provinces of Shaanxi, Ningxia, Qinghai, Inner Mongolia, and Sichuan.

Notes:

1. **Before answering the questions, be sure to fill in your name and exam number in the designated areas on the answer sheet.**
2. **When answering multiple-choice questions, use a 2B pencil to fill in the corresponding answer on the answer sheet. If you need to change your answer, erase the previous answer completely before filling in a new one.**
3. **When answering non-multiple-choice questions, use a 0.5mm black pen to write your answers in the designated areas on the answer sheet.**
4. **All answers must be written on the answer sheet. Answers written on the question paper will not be valid.**
5. **Only the answer sheet needs to be turned in after the exam.**

Section I: Multiple Choice Questions

This section contains 12 questions, each worth 5 points, for a total of 60 points. For each question, choose the one best answer from the four options provided.

1. 设 $z=5+i$，则 $i(\overline{z}+z)=($　　$)$

(A)10i　　(B)2i　　(C)10　　(D)−2

【参考答案】A

【详细解析】因为 $z=5+i$，所以 $i(\overline{z}+z)=10i$，故选(A).

Let z = 5 + i, then i(z + z) = (?)
(A) 10i (B) 2i (C) 10 (D) -2
Reference Answer: A
Detailed Explanation:
Because z = 5 + i, therefore $i(\overline{z}+z)=10i$, **so the answer is (A).**
In essence, this is a complex number problem where you are asked to calculate the value of i(z + z) given that z = 5 + i.

3. 若实数 x，y 满足约束条件(略)，则 $z=x-5y$ 的最小值为(　　)

(A)5　　(B)$\frac{1}{2}$　　(C)−2　　(D)$-\frac{7}{2}$

【参考答案】D

【详细解析】将约束条件两两联立可得 3 个交点：(0，−1)、$(\frac{3}{2}$，1)和(3，$\frac{1}{2}$)，经检验都符合约束条件．代入目标函数可得：$z_{\min}=-\frac{7}{2}$，故选(D).

Question 3:
If real numbers x and y satisfy the given constraints (omitted), then the minimum value of z = x - 5y is (?)
(A) 5 (B) 1/2 (C) -2 (D) -7/2
Reference Answer: D
Detailed Explanation:
By solving the system of constraints pairwise, we can obtain three intersection points: (0, -1), (3/2, 1), and (3, 2). After checking, all of them satisfy the constraints. Substituting into the objective function, we get: Zmin = -7/2, so the answer is (D).

4. 等差数列$\{a_n\}$的前 n 项和为 S_n，若 $S_5=S_{10}$，$a_5=1$，则 $a_1=($　　$)$

(A)−2　　(B)$\frac{7}{3}$　　(C)1　　(D)2

【参考答案】B

【详细解析】因为 $S_5=S_{10}$，所以 $S_7=S_{18}$，$a_8=0$，又因为 $a_5=1$，所以公差 $d=-\frac{1}{3}$，$a_1=a_8-7d=\frac{7}{3}$，故选(B).

Question 4:
Let {an} be an arithmetic sequence with Sn as the sum of its first n terms. If S5 = S10 and a5 = 1, then a1 = (?)
(A) -2 (B) 7/3 (C) 1 (D) 2
Reference Answer: B
Detailed Explanation:
Because S5 = S10, therefore S7 = S18, a8 = 0. Also, because a5 = 1, so the common difference d = -1/3, a1 = a5 - 7d = 7/3, hence the answer is (B).

5. 已知双曲线 C：$\frac{x^2}{a^2}-\frac{y^2}{b^2}=1(a>0，b>0)$的左、右焦点分别为 $F_1(0,$

(A)$\frac{13}{5}$ (B)$\frac{13}{7}$ (C)2 (D)3

【参考答案】C

【详细解析】$e=\frac{c}{a}=\frac{|F_1F_2|}{|PF_2|-|PF_1|}=2$，故选(C).

Question 5:
Given the hyperbola C: $x^2/a^2 - y^2/b^2 = 1$ ($a > 0, b > 0$) with left and right foci $F1(0, -c)$ and $F2(0, c)$ respectively, the value of $|F1F2|/a$ is (?)
(A) 13/5 (B) 13/7 (C) 2 (D) 3
Reference Answer: C
Detailed Explanation:
Since $e=\frac{c}{a}=\frac{|F_1F_2|}{|PF_2|-|PF_1|}=2$, **the answer is (C).**

6. 曲线 $f(x)=x^6+3x$ 在$(0，-1)$处的切线与坐标轴围成的面积为( )

(A)$\frac{1}{6}$ (B)$\frac{\sqrt{3}}{2}$ (C)$\frac{1}{2}$ (D)$\frac{\sqrt{3}}{2}$

【参考答案】A

【详细解析】因为 $y'=6x^5+3$，所以 $k=3$，$y=3x-1$，$S=\frac{1}{2}\times\frac{1}{3}\times1=\frac{1}{6}$，故选(A).

Question 6:
The area of the triangle formed by the tangent line to the curve $f(x)=x^6+3x$ at point (0, -1) and the coordinate axes is (?)
(A) 1/6 (B) √3/2 (C) 1/2 (D) √3/2
Reference Answer: A
Detailed Explanation:
Because $y'=6x^5+3$, **so k = 3, y = 3x - 1, S = 1/2 × 1 × 1/3 = 1/6, hence the answer is (A).**

8. 已知$\frac{\cos\alpha}{\cos\alpha-\sin\alpha}=\sqrt{3}$，则 $\tan(\alpha+\frac{\pi}{4})=$( )

(A)3 (B)$2\sqrt{3}-1$ (C)-3 (D)$\frac{1}{3}$

【参考答案】B

【详细解析】因为$\frac{\cos\alpha}{\cos\alpha-\sin\alpha}=\sqrt{3}$，所以 $\tan\alpha=1-\frac{\sqrt{3}}{3}$，$\tan(\alpha+\frac{\pi}{4})=\frac{\tan\alpha+1}{1-\tan\alpha}=2\sqrt{3}-1$，故选(B).

Question 8:
Given that: as $\frac{\cos\alpha}{\cos\alpha-\sin\alpha}=\sqrt{3}$, **Find the value of: tan(a + π/4) = (?)**
(A) 3 (B) 2√3 - 1 (C) -3 (D) 1/3
Reference Answer: B
Detailed Explanation:
Because: $\frac{\cos\alpha}{\cos\alpha-\sin\alpha}=\sqrt{3}$, **Therefore:** $\tan\alpha=1-\frac{\sqrt{3}}{3}$, $\tan(\alpha+\frac{\pi}{4})=\frac{\tan\alpha+1}{1-\tan\alpha}=2\sqrt{3}-1$, **Hence, the answer is (B).**

9. 已知向量 $\boldsymbol{a}=(x+1, x)$，$\boldsymbol{b}=(x, 2)$，则(　　)

(A)“$\boldsymbol{a}\perp\boldsymbol{b}$”的必要条件是“$x=-3$”　　(B)“$\boldsymbol{a}/\!/\boldsymbol{b}$”的必要条件是“$x=-3$”

(C)“$\boldsymbol{a}\perp\boldsymbol{b}$”的充分条件是“$x=0$”　　(D)“$\boldsymbol{a}/\!/\boldsymbol{b}$”的充分条件是“$x=0$”

【参考答案】C

【详细解析】$\boldsymbol{a}\perp\boldsymbol{b}$，则 $x(x+1)+2x=0$，解得：$x=0$ 或 -3，故选(C).

Question 9:
Given vectors a = (x+1, x), b = (x, 2), then ( )
(A) "a is perpendicular to b" is a necessary condition for "x = -3"
(B) "a is parallel to b" is a necessary condition for "x = -3"
(C) "a is perpendicular to b" is a sufficient condition for "x = 0"
(D) "a is parallel to b" is a sufficient condition for "x = 0"
Reference Answer: C
Detailed Explanation:
For a to be perpendicular to b, their dot product must be 0. So, x(x+1) + 2x = 0. Solving this equation, we get x = 0 or -3. Therefore, "a is perpendicular to b" is a sufficient condition for "x = 0". Hence, the answer is (C).

10. 已知已知 m、n 是两条不同的直线，α、β是两个不同的平面：①若 $m\perp\alpha$，$n\perp\alpha$，则 $m/\!/n$；②若$\alpha\cap\beta=m$，$m/\!/n$，则 $n/\!/\beta$；③若 $m/\!/\alpha$，$n/\!/\alpha$，m 与 n 可能异面，也可能相交，也可能平行；④若$\alpha\cap\beta=m$，n 与α和β所成的角相等，则 $m\perp n$，以上命题是真命题的是(　　)

(A)①③　　(B)②③　　(C)①②③　　(D)①③④

【参考答案】A

【详细解析】选(A).

Question 10:
Given that m and n are two different lines, and α and β are two different planes:
① If m is perpendicular to α and n is perpendicular to α, then m is parallel to n. ② If the intersection of α and β is m, and m is parallel to n, then n is parallel to β. ③ If m is parallel to α and n is parallel to α, then m and n could be skew lines, intersecting lines, or parallel lines. ④ If the intersection of α and β is m, and the angles formed by n and α are equal, then m and n are parallel lines.
Which of the above statements are true?
(A) 1, 3 (B) 2, 3 (C) 1, 2, 3 (D) 1, 3, 4
Reference Answer: A
Detailed Explanation:
Choose (A).

11. 在△ABC中，内角A，B，C所对边分别为a，b，c，若$B=\frac{\pi}{3}$，$b^2=\frac{9}{4}ac$，则$\sin A+\sin C=$(　　)

(A)$\frac{2\sqrt{39}}{13}$　　(B)$\frac{\sqrt{39}}{13}$　　(C)$\frac{\sqrt{7}}{2}$　　(D)$\frac{3\sqrt{13}}{13}$

【参考答案】C

【详细解析】因为$B=\frac{\pi}{3}$，$b^2=\frac{9}{4}ac$，所以$\sin A\sin C=\frac{4}{9}\sin^2B=\frac{1}{3}$．由余弦定理可得：$b^2=a^2+c^2-ac=\frac{9}{4}ac$，即：$a^2+c^2=\frac{13}{4}ac$，$\sin^2A+\sin^2C=\frac{13}{4}\sin A\sin C=\frac{13}{12}$，所以$(\sin A+\sin C)^2=\sin^2A+\sin^2C+2\sin A\sin C=\frac{7}{4}$，$\sin A+\sin C=\frac{\sqrt{7}}{2}$，故选(C).

Question 11:
In triangle ABC, sides opposite to angles A, B, and C are a, b, and c respectively. If B = π/3 and $b^2=\frac{9}{4}ac$, then sinA + sinC = (?)

(A)$\frac{2\sqrt{39}}{13}$　　(B)$\frac{\sqrt{39}}{13}$　　(C)$\frac{\sqrt{7}}{2}$　　(D)$\frac{3\sqrt{13}}{13}$

Reference Answer: C

Detailed Explanation:
Because $B=\frac{\pi}{3}$, $b^2=\frac{9}{4}ac$, so $\sin A\sin C=\frac{4}{9}\sin^2B=\frac{1}{3}$. By cosine law, we have: $b^2 = a^2 + c^2 - ac$ $=\frac{9}{4}ac$, which simplifies to $a^2 + c^2 = 13ac/4$, $\sin^2A+\sin^2C=\frac{13}{4}\sin A\sin C=\frac{13}{12}$. Therefore, $(\sin A+\sin C)^2=\sin^2A+\sin^2C+2\sin A\sin C=\frac{7}{4}$, $\sin A+\sin C=\frac{\sqrt{7}}{2}$, Therefore, the answer is (C).

12. 已知a，b，c成等差数列，直线$ax+by+c=0$与圆C：$x^2+(y+2)^2=5$交于A，B两点，则$|AB|$的最小值为(　　)

(A)2　　(B)3　　(C)4　　(D)6

【参考答案】C

【详细解析】因为a，b，c成等差数列，所以$a-2b+c=0$，直线$ax+by+c=0$恒过$P(1,-2)$，当$PC\perp AB$时，$|AB|$取得最小值，此时$|PC|=1$，$|AB|=2\sqrt{5-|PC|^2}=4$，故选(C).

Question 12:
Given that a, b, and c form an arithmetic sequence, the line ax + by + c = 0 intersects the circle C: $x^2 + (y+2)^2 = 5$ at points A and B. Then the minimum value of |AB| is (?)
(A) 2 (B) 3 (C) 4 (D) 6
Reference Answer: C
Detailed Explanation:
Because a, b, and c form an arithmetic sequence, we have a - 2b + c = 0. Therefore, the line ax + by + c = 0 always passes through point P(1, -2). When PC is perpendicular to AB, |AB| reaches its minimum value. At this time, |PC| = 1, and $|AB| = 2\sqrt{5 - |PC|^2} = 4$. Therefore, the answer is (C).

13. 二项式$(\frac{1}{3}+x)^{10}$的展开式中系数的最大值是________.

【参考答案】5

【详细解析】展开式中系数最大的项一定在下面的 5 项：$C_{10}^{5}(\frac{1}{3})^5$、$C_{10}^{6}(\frac{1}{3})^4$、$C_{10}^{7}(\frac{1}{3})^3$、$C_{10}^{8}(\frac{1}{3})^2$、$C_{10}^{9}(\frac{1}{3})^1$，计算可得：系数的最大值为$C_{10}^{8}(\frac{1}{3})^2=5$.

Question 13:
The largest coefficient in the expansion of (x + 1/3)^10 is _____.
Reference Answer: 5
Detailed Explanation:
The term with the largest coefficient in the expansion is definitely one of the following 5 terms: $C_{10}^{5}(\frac{1}{3})^5$、$C_{10}^{6}(\frac{1}{3})^4$、$C_{10}^{7}(\frac{1}{3})^3$、$C_{10}^{8}(\frac{1}{3})^2$、$C_{10}^{9}(\frac{1}{3})^1$, By calculating, we can get: the largest coefficient is $C_{10}^{8}(\frac{1}{3})^2=5$.

14. 甲、乙两个圆台上下底面的半径均为r_2和r_1，母线长分别为$2(r_1-r_2)$和$3(r_1-r_2)$，则两个圆台的体积之比$\frac{V_甲}{V_乙}=$________.

【参考答案】$\frac{\sqrt{6}}{4}$

【详细解析】$\frac{V_甲}{V_乙}=\frac{h_甲}{h_乙}=\frac{\sqrt{3}(r_1-r_2)}{2\sqrt{2}(r_1-r_2)}=\frac{\sqrt{6}}{4}$.

Question 14:
Two truncated cones, A and B, have upper and lower base radii of r_2 and r_1 respectively. Their slant heights are 2(r_1 - r_2) and 3(r_1 - r_2) respectively. The ratio of the volumes of the two truncated cones, Va/Vb, is equal to _____.
Reference Answer: $\frac{\sqrt{6}}{4}$
Detailed Explanation:
Va/Vb = $\frac{V_甲}{V_乙}=\frac{h_甲}{h_乙}=\frac{\sqrt{3}(r_1-r_2)}{2\sqrt{2}(r_1-r_2)}=\frac{\sqrt{6}}{4}$.

15. 已知$a>1$，$\frac{1}{\log_8 a}-\frac{1}{\log_a 4}=-\frac{5}{2}$，则$a=$__________.

【参考答案】64

【详细解析】因为$\frac{1}{\log_8 a}-\frac{1}{\log_a 4}=\frac{3}{\log_2 a}-\frac{1}{2}\log_2 a=-\frac{5}{2}$，所以$(\log_2 a+1)(\log_2 a-6)=0$，而$a>1$，故$\log_2 a=6$，$a=64$.

Question 15:
Given $a>1$, $\frac{1}{\log_8 a}-\frac{1}{\log_a 4}=-\frac{5}{2}$, find the value of a.
Reference Answer: 64
Detailed Explanation:
Because $\frac{1}{\log_8 a}-\frac{1}{\log_a 4}=\frac{3}{\log_2 a}-\frac{1}{2}\log_2 a=-\frac{5}{2}$, so $(\log_2 a + 1)(\log_2 a - 6) = 0$. Since $a > 1$, and therefore $\log_2 a = 6$, $a = 64$.

16. 编号为1、2、3、4、5、6的六个小球，不放回的抽取三次，记m表示前两个球号码的平均数，记n表示前三个球号码的平均数，则m与n差的绝对值不超过0.5的概率是_______.

【参考答案】$\frac{7}{15}$

【详细解析】记前三个球的号码分别为a、b、c，则共有$A_6^3=120$种可能．令$|m-n|=|\frac{a+b}{2}-\frac{a+b+c}{3}|=|\frac{a+b-2c}{6}|\leqslant 0.5$可得：$|a+b-2c|\leqslant 3$，根据对称性：$c=1$或6时，均有2种可能；$c=2$或5时，均有10种可能；$c=3$或4时，均有16种可能；故满足条件的共有56种可能，$P=\frac{56}{120}=\frac{7}{15}$.

Question 16:
There are six small balls numbered 1, 2, 3, 4, 5, and 6. Three balls are drawn without replacement. Let m represent the average of the numbers on the first two balls, and let n represent the average of the numbers on all three balls. The probability that the absolute value of the difference between m and n is no more than 0.5 is _____.

Reference Answer: 7/15

Detailed Explanation:

Let the numbers on the first three balls be a, b, and c, respectively. There are a total of $A_6^3= 120$ possible outcomes. Let $|m-n|=|\frac{a+b}{2}-\frac{a+b+c}{3}|=|\frac{a+b-2c}{6}|\leqslant 0.5$, which can be simplified to |a+b-2c| ≤ 3. Based on symmetry, when c = 1 or 6, there are 2 possibilities for each; when c = 2 or 5, there are 10 possibilities for each; when c = 3 or 4, there are 16 possibilities for each. Therefore, there are a total of 56 possible outcomes that satisfy the condition. The probability is P = 56/120 = 7/15.

18. (12分)(已知数列$\{a_n\}$的前n项和为S_n，且$4S_n=3a_n+4$.

(1)求$\{a_n\}$的通项公式；

(2)设$b_n=(-1)^{n-1}na_n$，求数列$\{b_n\}$的前n项和为T_n.

【参考答案】见解析.

【详细解析】(1)因为$4S_n=3a_n+4$，所以$4S_{n+1}=3a_{n+1}+4$，两式相减可得：$4a_{n+1}=3a_{n+1}-3a_n$，即：$a_{n+1}=-3a_n$，又因为$4S_1=3a_1+4$，所以$a_1=4$，故数列$\{a_n\}$是首项为4，公比为-3的等比数列，$a_n=4\cdot(-3)^{n-1}$；

(2)$b_n=(-1)^{n-1}na_n=4n\cdot 3^{n-1}$，所以$T_n=4(1\cdot 3^0+2\cdot 3^1+3\cdot 3^2+\cdots+n\cdot 3^{n-1})$，$3T_n=4(1\cdot 3^1+2\cdot 3^2+3\cdot 3^3+\cdots+n\cdot 3^n)$，两式相减可得：$-2T_n=4(1+3^1+3^2+\cdots+3^{n-1}-n\cdot 3^n)=4(\frac{1-3^n}{1-3}-n\cdot 3^n)=(2-4n)3^n-2$，$T_n=(2n-1)3^n+1$.

解法2：$b_n=(-1)^{n-1}na_n=4n\cdot 3^{n-1}$，所以$T_n=T_{n-1}+4n\cdot 3^{n-1}$，两边同时减去$(2n-1)3^n$可得：$T_n-(2n-1)3^n=T_{n-1}-(2n-3)3^{n-1}$，故$\{T_n-(2n-1)3^n\}$为常数列，即：$T_n-(2n-1)3^n=1$，$T_n=(2n-1)3^n+1$.

here's the English translation of the text in the image:

Question 18 (12 points)

Given that the sum of the first n terms of the sequence $\{a_n\}$ is S_n, and $4S_n = 3a_n + 4$.

(1) Find the general formula for $\{a_n\}$;

(2) Let $b_n = (-1)^{n-1}na_n$, Find the sum of the first n terms of the sequence $\{b_n\}$, denoted as T_n.

Reference Answer: See the detailed explanation.

Detailed Explanation:

(1) Because $4S_n = 3a_n + 4$, we have $4S_{n+1}=3a_{n+1}+4$, Subtracting the two equations, we get: $4a_{n+1}=3a_{n+1}-3a_n$, which simplifies to $a_{n+1}=-3a_n$, Since $4S_1=3a_1+4$, we have a1 = 4. Therefore, the sequence $\{a_n\}$ is a geometric sequence with the first term being 4 and the common ratio being -3. Thus, $a_n=4\cdot(-3)^{n-1}$; $3T_n=4(1\cdot3^1+2\cdot3^2+3\cdot3^3+\cdots+n\cdot3^n)$,

(2) $b_n=(-1)^{n-1}na_n=4n\cdot3^{n-1}$, so $T_n=4(1\cdot3^0+2\cdot3^1+3\cdot3^2+\cdots+n\cdot3^{n-1})$,

Subtracting the two equations, we get:

$-4n)3^n-2$, $T_n=(2n-1)3^n+1$.

$-2T_n=4(1+3^1+3^2+\cdots+3^{n-1}-n\cdot3^n)=4(\frac{1-3^n}{1-3}-n\cdot3^n)=(2$

Solution 2: $b_n=(-1)^{n-1}na_n=4n\cdot3^{n-1}$, so $T_n=T_{n-1}+4n\cdot3^{n-1}$, Subtracting (2n-1)*3^n from both sides, we get: $T_n-(2n-1)3^n=T_{n-1}-(2n-3)3^{n-1}$. Thus $\{T_n-(2n-1)3^n\}$ is a constant sequence, meaning $T_n-(2n-1)3^n=1$, $T_n=(2n-1)3^n+1$.

19．(12 分)如图，已知 $AB\,/\!/\,CD$，$CD\,/\!/\,EF$，$AB=DE=EF=CF=2$，$CD=4$，$AD=BC=\sqrt{10}$，$AE=2\sqrt{3}$，M 为 CD 的中点．

(1)证明：$EM\,/\!/$ 平面 BCF；

(2)求二面角 $A-EM-B$ 的正弦值．

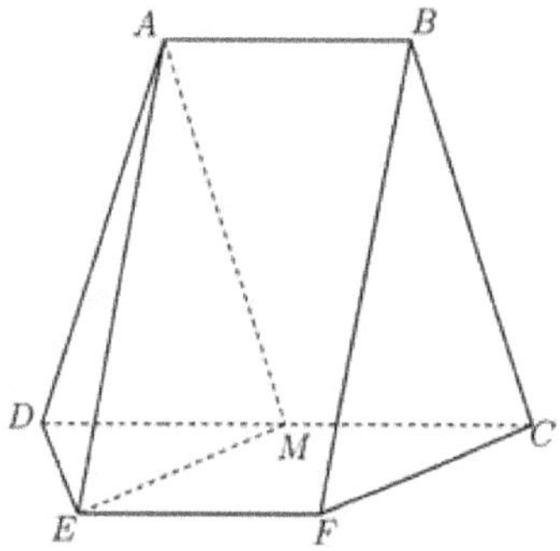

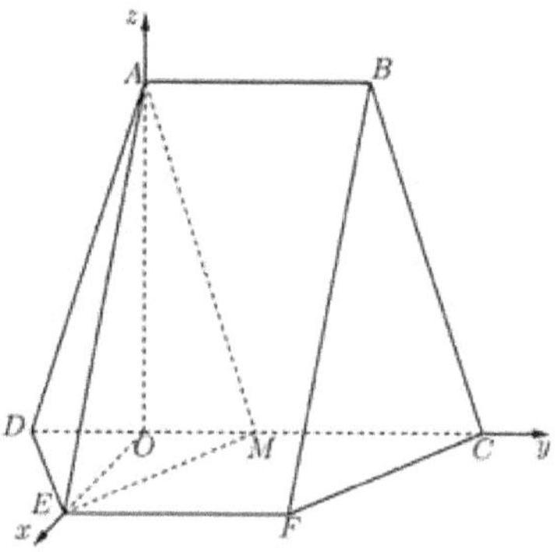

【详细解析】(1)由题意：$EF /\!/ CM$，$EF=CM$，而 $CF\subsetneq$平面 ADO，$EM\not\subset$平面 ADO，所以 EM $/\!/$ 平面 BCF；

(2)取 DM 的中点 O，连结 OA，OE，则 $OA\perp DM$，$OE\perp DM$，$OA=3$，$OE=\sqrt{3}$，而 $AE=2\sqrt{3}$，故 $OA\perp OE$．以 O 为坐标原点建立如图所示的空间直角坐标系，则 $A(0,\ 0,\ 3)$，$E(\sqrt{3},\ 0,\ 0)$，$M(0,\ 1,\ 0)$，$B(0,\ 2,\ 3)$，$\overrightarrow{AE}=(\sqrt{3},\ 0,\ -3)$，$\overrightarrow{EM}=(-\sqrt{3},\ 1,\ 0)$，$\overrightarrow{MB}=(0,\ 1,\ 3)$，

设平面 AEM 的法向量为$\vec{n}=(x,\ y,\ z)$，由$\begin{cases}\vec{n}\cdot\overrightarrow{AE}=0\\ \vec{n}\cdot\overrightarrow{EM}=0\end{cases}$可得：$\begin{cases}\sqrt{3}x-3z=0\\ -\sqrt{3}x+y=0\end{cases}$，令 $z=1$，则$\vec{n}=(\sqrt{3},\ 3,\ 1)$，同理：取平面 BEM 的法向量为$\vec{m}=(\sqrt{3},\ 3,\ -1)$，则 $\cos<\vec{m},\ \vec{n}>=\dfrac{\vec{m}\cdot\vec{n}}{|\vec{m}||\vec{n}|}=\dfrac{11}{13}$，$\sin<\vec{m},\ \vec{n}>=\dfrac{4\sqrt{3}}{13}$，故二面角 $A-EM-B$ 的正弦值为$\dfrac{4\sqrt{3}}{13}$.

Question 19 (12 points)

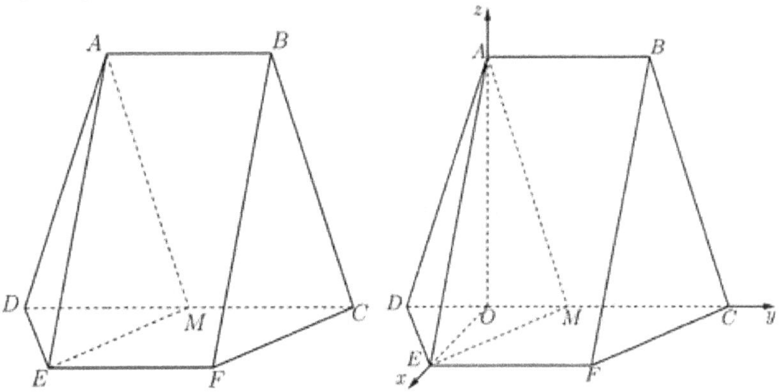

As shown in the figure, given that AB // CD, CD // EF, AB = DE = EF = CF = 2, CD = 4, AD = BC = √10, AE = 2√3, M is the midpoint of CD.

(1) Prove: EM // plane BCF;

(2) Find the sine value of the dihedral angle A-EM-B.

Detailed Analysis:

(1) Based on the given information: EF // CM, EF = CM, and CF is on plane ADO, EM is on plane ADO, therefore EM // plane BCF.

(2) Take the midpoint of DM as point O. Connect OA and OE. Then

$OA\perp DM$，$OE\perp DM$，$OA=3$，$OE=\sqrt{3}$，

Therefore, OA ⊥ OE. Taking O as the origin, establish a spatial rectangular coordinate system as shown in the figure. Then A(0, 0, 3), E(√3, 0, 0), M(0, 1, 0), B(0, 2, 3)

$\overrightarrow{AE}=(\sqrt{3},\ 0,\ -3)$，$\overrightarrow{EM}=(-\sqrt{3},\ 1,\ 0)$，$\overrightarrow{MB}=(0,\ 1,$ Let the normal vector of plane AEM be $\vec{n}=(x,\ y,\ z)$, from $\begin{cases}\vec{n}\cdot\overrightarrow{AE}=0\\ \vec{n}\cdot\overrightarrow{EM}=0\end{cases}$, we can get $\begin{cases}\sqrt{3}x-3z=0\\ -\sqrt{3}x+y=0\end{cases}$ Let z = 1, then

$\vec{n}=(\sqrt{3}, 3, 1)$, Similarly, let the normal vector of plane BEM be $\vec{m}=(\sqrt{3}, 3, -1)$, then $\cos<\vec{m}, \vec{n}>=\frac{\vec{m}\cdot\vec{n}}{|\vec{m}||\vec{n}|}=\frac{11}{13}$, $\sin<\vec{m}, \vec{n}>=\frac{4\sqrt{3}}{13}$, so the sine of the dihedral angle A-EM-B is $\frac{4\sqrt{3}}{13}$.

20．(12 分)已知函数 $f(x)=(1-ax)\ln(1+x)-x$.

(1)当 $a=-2$ 时，求 $f(x)$的极值；

(2)当 $x\geq 0$ 时，$f(x)\geq 0$，求 a 的取值范围.

【详细解析】(1)当 $a=-2$ 时，$f(x)=(1+2x)\ln(1+x)-x$，$x>-1$．$f'(x)=2\ln(1+x)+\frac{x}{1+x}$，当 $x>0$ 时，$f'(x)>0$，当 $-1<x<0$ 时，$f'(x)<0$，所以 $f(x)$在$(-1, 0)$上递减，在$(0, +\infty)$上递增，故 $f(x)$的极小值为 $f(0)=0$，无极大值；

(2)$f(x)=(1-ax)\ln(1+x)-x$，$f'(x)=-a\ln(1+x)-\frac{(a+1)x}{1+x}$．令 $g(x)=f'(x)$，则 $g'(x)=-\frac{a}{1+x}-\frac{a+1}{(1+x)^2}$．因为当 $x\geq 0$ 时，$f(x)\geq 0$，且 $f(0)=0$，$f'(0)=0$，所以 $g'(0)=-1-2a\geq 0$，$a\leq -\frac{1}{2}$.

当 $a\leq -\frac{1}{2}$时，$g'(x)\geq \frac{1}{2(1+x)}-\frac{1}{2(1+x)^2}=\frac{x}{2(1+x)^2}\geq 0$，$g(x)$在$[0, +\infty)$上递增，$g(x)=f'(x)\geq g(0)=0$，故 $f(x)$在$[0, +\infty)$上递增，$f(x)\geq f(0)=0$ 恒成立，即 a 的取值范围为$(-\infty, -\frac{1}{2}]$.

Question 20 (12 points)
Given the function f(x) = (1 - ax)ln(1 + x) - x.
(1) When a = -2, find the extreme values of f(x).
(2) When x ≥ 0, f(x) ≥ 0, find the range of values for a.

Detailed Explanation:
(1) When a = -2, f(x) = (1 + 2x)ln(1 + x) - x, x > -1. f'(x) = 2ln(1 + x) + x/(1 + x).
When x > 0, f'(x) > 0; when -1 < x < 0, f'(x) < 0. Therefore, f(x) decreases on (-1, 0) and increases on (0, +∞). Thus, the minimum value of f(x) is f(0) = 0, and there is no maximum value.
(2) f(x) = (1 - ax)ln(1 + x) - x, f'(x) = -aln(1 + x) - (a + 1)x/(1+). Let g(x) = f'(x), then g'(x) = -a/(1 + x) - (a + 1)/(1 + x)^2.

Because when x ≥ 0, f(x) ≥ 0, and f(0) = 0, f'(0) = 0, so $g'(0)=-1-2a\geq 0$, $a\leq -\frac{1}{2}$.
When a ≤ -1/2, $g'(x)\geq \frac{1}{2(1+x)}-\frac{1}{2(1+x)^2}=\frac{x}{2(1+x)^2}\geq 0$, g(x) is increasing on [0, +∞), g(x) = f'(x) ≥ g(0) = 0, so f(x) is increasing on [0, +∞), f(x) > f(0) = 0 always holds, that is, the range of values for a is (-∞, -1/2].

21．(12 分)(已知椭圆 C：$\frac{x^2}{a^2}+\frac{y^2}{b^2}=1(a>b>0)$的右焦点为 F，点 $M(1，\frac{3}{2})$在椭圆 C 上，且 $MF\perp x$ 轴．

(1)求椭圆 C 的方程；

(2)$P(4，0)$，过 P 的直线与椭圆 C 交于 A， B 两点，N 为 FP 的中点，直线 NB 与 MF 交于 Q，证明：$AQ\perp y$ 轴．

【参考答案】见解析

【详细解析】(1)设椭圆 C 的左焦点为 F_1，则$|F_1F|=2$，$|MF|=\frac{3}{2}$．因为 $MF\perp x$ 轴，所以$|MF_1|=\frac{5}{2}$，$2a=|MF_1|+|MF|=4$，解得：$a^2=4$，$b^2=a^2-1=3$，故椭圆 C 的方程为：$\frac{x^2}{4}+\frac{y^2}{3}=1$；

(2)解法 1：设 $A(x_1，y_1)$，$B(x_2，y_2)$，$\overrightarrow{AP}=\lambda\overrightarrow{PB}$，则$\begin{cases}\frac{x_1+\lambda x_2}{1+\lambda}=4\\ \frac{y_1+\lambda y_2}{1+\lambda}=0\end{cases}$，即$\begin{cases}\lambda x_2=4+4\lambda-x_1\\ \lambda y_2=-y_1\end{cases}$．又由$\begin{cases}3x_1^2+4y_1^2=12\\ 3(\lambda x_2)^2+4(\lambda y_2)^2=12\lambda^2\end{cases}$可得：$3\cdot\frac{x_1+\lambda x_2}{1+\lambda}\cdot\frac{x_1-\lambda x_2}{1-\lambda}+4\cdot\frac{y_1+\lambda y_2}{1+\lambda}\cdot\frac{y_1-\lambda y_2}{1-\lambda}=12$，结合上式可得：$5\lambda-2\lambda x_2+3=0$．$P(4，0)$，$F(1，0)$，$N(\frac{5}{2}，0)$，则 $y_Q=\frac{3y_2}{5-2x_2}=\frac{3\lambda y_2}{5\lambda-2\lambda x_2}=-\lambda y_2=y_1$，故 $AQ\perp y$ 轴．

解法 2：设 $A(x_1，y_1)$，$B(x_2，y_2)$，，则$\frac{y_1}{x_1-4}=\frac{y_2}{x_2-4}$，即：$x_1y_2-x_2y_1=4(y_2-y_1)$，所以$(x_1y_2-x_2y_1)(x_1y_2+x_2y_1)=x_1^2y_2^2-x_2^2y_1^2=(4+\frac{4y_1^2}{3})y_2^2-(4+\frac{4y_2^2}{3})y_1^2=4(y_2-y_1)(y_2+y_1)=4(y_2-y_1)(x_1y_2+x_2y_1)$，即：$x_1y_2+x_2y_1=y_2+y_1$，$2x_2y_1=5y_1-3y_2$．$P(4，0)$，$F(1，0)$，$N(\frac{5}{2}，0)$，则 $y_Q=\frac{3y_2}{5-2x_2}=\frac{3y_1y_2}{5y_1-2y_1x_2}=y_1$，故 $AQ\perp y$ 轴．

Question 21 (12 points)

Given the ellipse C: $x^2/a^2 + y^2/b^2 = 1$ ($a > b > 0$) with right focus F, point M(1, 3/2) is on the ellipse C, and MF is perpendicular to the x-axis.

(1) Find the equation of ellipse C;

(2) P(4, 0), a straight line passing through P intersects the ellipse C at points A and B, N is the midpoint of FP, line NB intersects MF at point Q, prove: AQ is perpendicular to the y-axis.

Reference Answer: See detailed explanation.

Detailed Explanation:

(1) Let the left focus of ellipse C be F_1, then $|FF_1| = 2$, $|MF| = 3/2$, because MF is perpendicular to the x-axis, so $|MF_1| = 5/2$, $2a = |MF| + |MF_1| = 4$, solving for $a^2 = 4$, $b^2 = a^2 - c^2-1 = 3$, so the equation of ellipse C is $x^2/4 + y^2/3 = 1$;

(2) Solution 1: Let A(x₁, y₁), B(x₂, y₂), $\overrightarrow{AP}=\lambda\overrightarrow{PB}$, then $\begin{cases}\dfrac{x_1+\lambda x_2}{1+\lambda}=4\\ \dfrac{y_1+\lambda y_2}{1+\lambda}=0\end{cases}$, which simplifies to $\begin{cases}\lambda x_2=4+4\lambda-x_1\\ \lambda y_2=-y_1\end{cases}$. and from $\begin{cases}3x_1^2+4y_1^2=12\\ 3(\lambda x_2)^2+4(\lambda y_2)^2=12\lambda^2\end{cases}$ we can get: $3\cdot\dfrac{x_1+\lambda x_2}{1+\lambda}\cdot\dfrac{x_1-\lambda x_2}{1-\lambda}+4\cdot\dfrac{y_1+\lambda y_2}{1+\lambda}\cdot\dfrac{y_1-\lambda y_2}{1-\lambda}=12$, , combining the above equation, we get: $5\lambda-22x_2\lambda+3=0$. P(4, 0), F(1, 0), N(5/2, 0), $y_Q=\dfrac{3y_2}{5-2x_2}=\dfrac{3\lambda y_2}{5\lambda-2\lambda x_2}=-\lambda y_2=y_1$, so AQ is perpendicular to the y-axis.

Solution 2: Let A(x₁, y₁), B(x₂, y₂), then $\dfrac{y_1}{x_1-4}=\dfrac{y_2}{x_2-4}$, so $x_1y_2-x_2y_1=4(y_2-y_1)$, so $(x_1y_2-x_2y_1)(x_1y_2+x_2y_1)=x_1^2y_2^2-x_2^2y_1^2=(4+\dfrac{4y_1^2}{3})y_2^2-(4+\dfrac{4y_2^2}{3})y_1^2=4(y_2-y_1)(y_2+y_1)=4(y_2-y_1)(x_1y_2+x_2y_1)$, that is $x_1y_2+x_2y_1=y_2+y_1$, $2x_2y_1=5y_1-3y_2$. $P(4, 0)$, $F(1, 0)$, $N(\dfrac{5}{2}, 0)$, then $y_Q=\dfrac{3y_2}{5-2x_2}=\dfrac{3y_1y_2}{5y_1-2y_1x_2}=y_1$, so AQ is perpendicular to the y-axis.

22．[选修 4－4：坐标系与参数方程](10 分)( )在平面直角坐标系 xOy 中，以坐标原点 O 为极点，x 轴的正半轴为极轴建立极坐标系，曲线 C 的极坐标方程为 $\rho=\rho\cos\theta+1$.

(1)写出 C 的直角坐标方程；

(2)直线 $\begin{cases}x=t\\ y=t+a\end{cases}$ (t 为参数)与曲线 C 交于 A、B 两点，若 $|AB|=2$，求 a 的值.

【参考答案】见解析

【详细解析】(1)因为 $\rho=\rho\cos\theta+1$，所以 $\rho^2=(\rho\cos\theta+1)^2$，故 C 的直角坐标方程为：$x^2+y^2=(x+1)^2$，即：$y^2=2x+1$；

(2)将 $\begin{cases}x=t\\ y=t+a\end{cases}$ 代入 $y^2=2x+1$ 可得：$t^2+2(a-1)t+a^2-1=0$，$|AB|=\sqrt{2}|t_1-t_2|=\sqrt{16(1-a)}=2$，

解得：$a=\dfrac{3}{4}$.

Question 22 (10 points)

In the rectangular coordinate system xOy, with the origin as the pole and the positive x-axis as the polar axis, establish a polar coordinate system. The polar equation of curve C is $\rho=\rho\cos\theta+1$.

(1) Write the rectangular coordinate equation of C;

(2) The straight line $\begin{cases} x=t \\ y=t+a \end{cases}$ (t is a parameter) intersects curve C at points A and B. If |AB| = 2, find the value of a.

Reference Answer: See detailed explanation.

Detailed Explanation:

(1) Because $\rho = \rho\cos\theta + 1$, so $\rho^2 = (\rho\cos\theta + 1)^2$, thus the rectangular coordinate equation of C is: $x^2 + y^2 = (x + 1)^2$, that is: $y^2 = 2x + 1$;

(2) Substituting $\begin{cases} x=t \\ y=t+a \end{cases}$ into $y^2 = 2x + 1$, we get:

$t^2+2(a-1)t+a^2-1=0$，$|AB|=\sqrt{2}|t_1-t_2|=\sqrt{16(1-a)}=2$，, solving for a: a = 3/4.

23．[选修4－5：不等式选讲](10分)(　　　　　　　　　　)实数 a，b 满足 $a+b\geq3$.

(1)证明：$2a^2+2b^2>a+b$；

(2)证明：$|a-2b^2|+|b-2a^2|\geq6$.

【解析】(1)因为 $a+b\geq3$，所以 $2a^2+2b^2\geq(a+b)^2>a+b$；

Question 23 (10 points)

Real numbers a and b satisfy $a + b \geq 3$.

(1) Prove: $2a^2 + 2b^2 > a + b$;

(2) Prove: $|a - 2b^2| + |b - 2a^2| \geq 6$.

Solution:

(1) Because $a + b \geq 3$, so $2a^2 + 2b^2 = (a + b)^2 > a + b$;

(2)$|a-2b^2|+|b-2a^2|\geq|a-2b^2+b-2a^2|=|2a^2+2b^2-(a+b)|=2a^2+2b^2-(a+b)\geq(a+b)^2-(a+b)=(a+b)(a+b-1)\geq6$.

1．(　　　　　　　　　　)集合 $A=\{1，2，3，4，5，9\}$，$B=\{x|x+1\in A\}$，则 $A\cap B=$(　　)

(A){1，2，3，4}　(B){1，2，3，4}　(C){1，2，3，4}　(D){1，2，3，4}

【参考答案】A

【详细解析】因为 $A=\{1，2，3，4，5，9\}$，$B=\{x|x+1\in A\}=\{0，1，2，3，4，8\}$，所以 $A\cap B=\{1，2，3，4\}$，故选(A).

Let A={1,2,3,4,5,9} and B={x|x+1∈A}. Then A∩B=()

(A){1,2,3,4}(B){1,2,3,4}(C){1,2,3,4}(D){1,2,3,4}

【Reference answer】A

【Detailed analysis】Because A={1,2,3,4,5,9}, B={x|x+1∈A}={0,1,2,3,4,8}, so A∩B={1,2,3,4}, so the answer is (A).

2．(　　　　　　　　　　)设 $z=\sqrt{2}i$，则 $z\cdot\bar{z}=$(　　)

(A)2　(B)2　(C)2　(D)2

【参考答案】D

【详细解析】因为 $z=\sqrt{2}i$，所以 $z\cdot\bar{z}=2$，故选(D).

Let $z=\sqrt{2}i$, then $z\cdot\bar{z}=$ ()

(A) 2 (B) 2 (C) 2 (D) 2

【Reference answer】D

【Detailed analysis】Because z = $\sqrt{2}$i, so $z\cdot\bar{z}$ = 2, hence choose (D).

3. ( )若实数 x，y 满足约束条件(略)，则 $z=x-5y$ 的最小值为( )

(A)5 (B)$\frac{1}{2}$ (C)-2 (D)$-\frac{7}{2}$

【参考答案】D

【详细解析】将约束条件两两联立可得 3 个交点：(0，-1)、($\frac{3}{2}$，1)和(3，$\frac{1}{2}$)，经检验都符合约束条件．代入目标函数可得：$z_{\min}=-\frac{7}{2}$，故选(D).

Question 3

If real numbers x and y satisfy the constraint conditions (omitted), then the minimum value of z = x - 5y is ( ).

(A) 5 (B) 1/2 (C) -2 (D) -7/2

Reference Answer: D

Detailed Explanation: Solving the constraint conditions in pairs, we get three intersection points: (0, -1), (3/2, 1), and (3, 1/2). After checking, all of them satisfy the constraint conditions. Substituting into the objective function, we get: Zmin = -7/2, so the answer is (D).

4. ( )等差数列$\{a_n\}$的前 n 项和为 S_n，若 $S_9=1$，$a_3+a_7=$( )

(A)-2 (B)$\frac{7}{3}$ (C)1 (D)$\frac{2}{9}$

【参考答案】D

【详细解析】令 $d=0$，则 $S_9=9a_n=1$，$a_n=\frac{1}{9}$，$a_3+a_7=\frac{2}{9}$，故选(D).

Question 4

Let Sn be the sum of the first n terms of the arithmetic sequence {an}. If S9=1, then $a_3 + a_7 = $ ( )

(A) -2 (B) 7/3 (C) 1 (D) 2/9

Reference Answer: D

Detailed Explanation: Let d = 0, then $S_9=9a_n=1$，$a_n=\frac{1}{9}$，$a_3+a_7=\frac{2}{9}$，
Therefore, the answer is (D).

5. ( )甲、乙、丙、丁四人排成一列，丙不在排头，且甲或乙在排尾的概率是( )

(A)$\frac{1}{4}$ (B)$\frac{1}{3}$ (C)$\frac{1}{2}$ (D)$\frac{2}{3}$

【参考答案】B

【详细解析】甲、乙、丙、丁四人排成一列共有 24 种可能．丙不在排头，且甲或乙在排尾的共有 8 种可能，$P=\frac{8}{24}=\frac{1}{3}$，故选(B).

Question 5

There are four people: A, B, C, and D. They are arranged in a row. C is not at the head of the row, and either A or B is at the end of the row. The probability of this arrangement is ( ).

(A) 1/4 (B) 1/3 (C) 1/2 (D) 2/3

Reference Answer: B

Detailed Explanation: There are 24 possible ways to arrange the four people A, B, C, and D in a row. There are 8 possible ways to arrange them such that C is not at the head of the row and either A or B is at the end of the row. Therefore, the probability is 8/24 = 1/3, so the answer is (B).

6．(　　　　　　　　　　　)已知双曲线 C：$\frac{x^2}{a^2}-\frac{y^2}{b^2}=1(a>0，b>0)$的左、右焦点分别为 $F_1(0，4)$、$F_2(0，-4)$，且经过点 $P(-6，4)$，则双曲线 C 的离心率是(　　)

(A)$\frac{13}{5}$　　(B)$\frac{13}{7}$　　(C)2　　(D)3

【参考答案】C

【详细解析】$e=\frac{c}{a}=\frac{|F_1F_2|}{|PF_2|-|PF_1|}=2$，故选(C).

Question 6

Given the hyperbola C：$\frac{x^2}{a^2}-\frac{y^2}{b^2}=1(a>0，b>0)$ with left and right foci F1(0, 4) and F2(0, -4), and passing through point P(-6, 4), then the eccentricity of hyperbola C is ( ).

(A)$\frac{13}{5}$　　(B)$\frac{13}{7}$　　(C)2　　(D)3

Reference Answer: C

Detailed Explanation: $e=\frac{c}{a}=\frac{|F_1F_2|}{|PF_2|-|PF_1|}=2$, so choose (C).

7.(　　　　　　　　　　　)曲线 $f(x)=x^6+3x$ 在$(0，-1)$处的切线与坐标轴围成的面积为(　　)

(A)$\frac{1}{6}$　　(B)$\frac{\sqrt{3}}{2}$　　(C)$\frac{1}{2}$　　(D)$\frac{\sqrt{3}}{2}$

【参考答案】A

【详细解析】因为 $y'=6x^5+3$，所以 $k=3$，$y=3x-1$，$S=\frac{1}{2}\times\frac{1}{3}\times 1=\frac{1}{6}$，故选(A).

Question 7.

The area enclosed by the tangent line to the curve f(x) = x^3 + 3x at point (0, -1) and the coordinate axes is ( ).

(A) 1/6 (B) √3/2 (C) 1/2 (D) √3/2

Reference Answer: A

Detailed Explanation: Because y' = 6x + 3, so k = 3, y = 3x - 1, S = (1/2) * 1 * 1/3 = 1/6. Therefore, choose (A).

9．(　　　　　　　　　　　)已知$\frac{\cos\alpha}{\cos\alpha-\sin\alpha}=\sqrt{3}$，则 $\tan(\alpha+\frac{\pi}{4})=$(　　)

(A)3　　(B)$2\sqrt{3}-1$　　(C)-3　　(D)$\frac{1}{3}$

【参考答案】B

【详细解析】因为$\frac{\cos\alpha}{\cos\alpha-\sin\alpha}=\sqrt{3}$，所以 $\tan\alpha=1-\frac{\sqrt{3}}{3}$，$\tan(\alpha+\frac{\pi}{4})=\frac{\tan\alpha+1}{1-\tan\alpha}=2\sqrt{3}-1$，故选(B).

Question 9.

Given that cosa−sinacosa=3　　, find $\tan(a + \frac{\pi}{4}) = $ ( )

(A) 3 (B) 23　　−1 (C) -3 (D) 31

Reference Answer: B

Detailed Explanation: Because $\frac{\cos\alpha}{\cos\alpha-\sin\alpha}=\sqrt{3}$, so $\tan\alpha=1-\frac{\sqrt{3}}{3}$, $\tan(\alpha+\frac{\pi}{4})=\frac{\tan\alpha+1}{1-\tan\alpha}=2\sqrt{3}-1$, hence choose (B).

12. ，在△ABC中，内角A，B，C所对边分别为a，b，c，若$B=\frac{\pi}{3}$，$b^2=\frac{9}{4}ac$，则$\sin A+\sin C=$(　　)

(A) $\frac{2\sqrt{39}}{13}$　　(B) $\frac{\sqrt{39}}{13}$　　(C) $\frac{\sqrt{7}}{2}$　　(D) $\frac{3\sqrt{13}}{13}$

【参考答案】C

【详细解析】因为$B=\frac{\pi}{3}$，$b^2=\frac{9}{4}ac$，所以$\sin A\sin C=\frac{4}{9}\sin^2B=\frac{1}{3}$．由余弦定理可得：$b^2=a^2+c^2-ac=\frac{9}{4}ac$，即：$a^2+c^2=\frac{13}{4}ac$，$\sin^2A+\sin^2C=\frac{13}{4}\sin A\sin C=\frac{13}{12}$，所以$(\sin A+\sin C)^2=\sin^2A+\sin^2C+2\sin A\sin C=\frac{7}{4}$，$\sin A+\sin C=\frac{\sqrt{7}}{2}$，故选(C).

Question 12.

In triangle ABC, the sides opposite angles A, B, and C are a, b, and c respectively. If B = π/3 and b^2 = (9/4)ac, then sinA + sinC = ( )

(A) (2√39)/13 (B) √39/13 (C) √7/2 (D) √13/3

Reference Answer: C

Detailed Explanation: Because B = π/3 and b^2 = (9/4)ac, so sinAsinC = $\sin^2$B = 1/3. From the cosine law, we have: $b^2=a^2+c^2-ac=\frac{9}{4}ac$, which implies $a^2+c^2=\frac{13}{4}ac$, $\sin^2A+\sin^2C=\frac{13}{4}\sin A\sin C=\frac{13}{12}$, Therefore, $(\sin A+\sin C)^2=\sin^2A+\sin^2C+2\sin A\sin C=\frac{7}{4}$, $\sin A+\sin C=\frac{\sqrt{7}}{2}$, hence choose (C).

14. (　　)函数$f(x)=\sin x-\sqrt{3}\cos x$在$[0，\pi]$上的最大值是________.

【参考答案】2

【详细解析】$f(x)=\sin x-\sqrt{3}\cos x=2\sin(x-\frac{\pi}{3})\leqslant 2$，当且仅当$x=\frac{5\pi}{6}$时取等号.

Question 14.

The maximum value of the function $f(x)=\sin x-\sqrt{3}\cos x$ on the interval [0, π] is _____.

Detailed Explanation: $f(x)=\sin x-\sqrt{3}\cos x$ = 2sin(x - π/3) ≤ 2, with equality holding if and only if x = 5π/6.

15. (　　)已知$a>1$，$\frac{1}{\log_8 a}-\frac{1}{\log_a 4}=-\frac{5}{2}$，则$a=$__________.

【参考答案】64

【详细解析】因为$\frac{1}{\log_8 a}-\frac{1}{\log_a 4}=\frac{3}{\log_2 a}-\frac{1}{2}\log_2 a=-\frac{5}{2}$，所以$(\log_2 a+1)(\log_2 a-6)=0$，而$a>1$，故$\log_2 a=6$，$a=64$.

Given $a>1$, $\frac{1}{\log_8 a}-\frac{1}{\log_a 4}=-\frac{5}{2}$, find a. **Reference Answer:** 64

Detailed Explanation: Because $\frac{1}{\log_8 a}-\frac{1}{\log_a 4}=\frac{3}{\log_2 a}-\frac{1}{2}\log_2 a=-\frac{5}{2}$, so $(\log_2 a+1)(\log_2 a-6)=0$, and since a>1, therefore, $\log_2 a=6$, $a=64$.

16．(   )曲线 $y=x^3-3x$ 与 $y=-(x-1)^2+a$ 在$(0，+\infty)$上有两个不同的交点，则 a 的取值范围为_______.

【参考答案】$(-2，1)$

【详细解析】令 $x^3-3x=-(x-1)^2+a$，则 $a=x^3-3x+(x-1)^2$，设$\varphi(x)=x^3-3x+(x-1)^2$，$\varphi'(x)=(3x+5)(x-1)$，$\varphi(x)$在$(1，+\infty)$上递增，在$(0，1)$上递减．因为曲线 $y=x^3-3x$ 与 $y=-(x-1)^2+a$ 在$(0，+\infty)$上有两个不同的交点，$\varphi(0)=1$，$\varphi(1)=-2$，所以 a 的取值范围为$(-2，1)$.

Question 16.
The curves $y = x^3 - 3x$ and $y = (x - 1)^2 + a$ have two different intersection points in the interval $(0, +\infty)$. The range of values for a is _____.
Reference Answer: (-2, 1)
Detailed Explanation: Let $x^3 - 3x =- (x - 1)^2 + a$, then $a = x^3 - 3x + (x - 1)^2$. Let $\varphi(x) = x^3 - 3x + (x - 1)^2$, $\varphi'(x) = (3x + 5)(x - 1)$. $\varphi(x)$ is increasing on $(1, +\infty)$ and decreasing on $(0, 1)$. Because the curves $y = x^3 - 3x$ and $y = (x - 1)^2 + a$ have two different intersection points in $(0, +\infty)$, $\varphi(0) = 1$, $\varphi(1) = -2$, so the range of values for a is (-2, 1).

17．(12 分)(   )已知等比数列$\{a_n\}$的前 n 项和为 S_n，且 $2S_n=3a_{n+1}-3$.

(1)求$\{a_n\}$的通项公式；

(2)求数列$\{S_n\}$的通项公式.

【参考答案】见解析.

【详细解析】(1)因为 $2S_n=3a_{n+1}-3$，所以 $2S_{n+1}=3a_{n+2}-3$，两式相减可得：$2a_{n+1}=3a_{n+2}-3a_{n+1}$，即：$3a_{n+2}=5a_{n+1}$，所以等比数列$\{a_n\}$的公比 $q=\frac{5}{3}$，又因为 $2S_1=3a_2-3=5a_1-3$，所以 $a_1=1$，$a_n=(\frac{5}{3})^{n-1}$；

(2)因为 $2S_n=3a_{n+1}-3$，所以 $S_n=\frac{3}{2}(a_{n+1}-1)=\frac{3}{2}[(\frac{5}{3})^n-1]$.

Question 17. (12 points)
Given that the sum of the first n terms of the geometric sequence $\{a_n\}$ is Sn, and 2Sn = $3a_n$+1 - 3.
(1) Find the general formula for $\{a_n\}$;
(2) Find the general formula for the sequence {Sn}.
Reference Answer: See detailed explanation.
Detailed Explanation:
(1) Because $2S_n=3a_{n+1}-3$, so $2S_{n+1}=3a_{n+2}-3$, Subtracting the two equations, we get: $2a_{n+1}=3a_{n+2}-3a_{n+1}$, so $3a_{n+2}=5a_{n+1}$, Therefore, the common ratio of the geometric sequence

$\{a_n\}$ is q = 5/3. Also, because $2S_1 = 3a_2 - 3 = 5a_1 - 3$, so $a_1 = 1$, $a_n = (\frac{5}{3})^{n-1}$;

(2) Because $2S_n = 3a_{n+1} - 3$, so $S_n = \frac{3}{2}(a_{n+1} - 1) = \frac{3}{2}[(\frac{5}{3})^n - 1]$.

2023 Beijing College Entrance Exam
Mathematics
This exam is worth 150 points and lasts 120 minutes. All answers must be written on the answer sheet. After the exam, both the exam paper and the answer sheet must be returned.
Section 1: Multiple Choice
This section has 10 questions, each worth 4 points, for a total of 40 points. Choose the best answer for each question from the four options provided.

1. Given sets $M = \{x \mid x + 2 \geq 0\}$, $N = \{x \mid x - 1 < 0\}$, find $M \cap N$
A. $\{x \mid -2 \leq x < 1\}$
B. $\{x \mid -2 < x \leq 1\}$
C. $\{x \mid x \geq -2\}$
D. $\{x \mid x < 1\}$

1. Answer: A
Analysis: First simplify the sets M and N, then calculate their intersection based on the definition of intersection.
Detailed explanation:
From the question, $M = \{x \mid x + 2 \geq 0\} = \{x \mid x \geq -2\}$, $N = \{x \mid x - 1 < 0\} = \{x \mid x < 1\}$.
According to the definition of intersection, $M \cap N = \{x \mid -2 \leq x < 1\}$.
Therefore, the answer is A.

2. On the complex plane, if the coordinates of the point corresponding to complex number z are $(-1, \sqrt{3})$, then the conjugate of z is ()

A. $1+\sqrt{3}i$ B. $1-\sqrt{3}i$

C. $-1+\sqrt{3}i$ D. $-1-\sqrt{3}i$

2. Answer: D
Analysis: Based on the geometric meaning of complex numbers, we first find the complex number z, and then use the definition of the conjugate of a complex number to calculate.
Detailed explanation:
On the complex plane, the corresponding point is $(-1, \sqrt{3})$. According to the geometric meaning of complex numbers, $z = -1 + \sqrt{3}i$.
From the definition of the conjugate of a complex number, we have $\bar{z} = -1 - \sqrt{3}i$.
Therefore, the answer is D.

3. **Given vectors $\vec{a}$, $\vec{b}$, satisfying $\vec{a}+\vec{b}=(2,3), \vec{a}-\vec{b}=(-2,1)$, find $|\vec{a}|^2-|\vec{b}|^2$**

- A. -2
- B. -1
- C. 0
- D. 1

3. Answer: B
Analysis: Utilize the operation laws of the dot product of plane vectors and the coordinate representation of the dot product to solve the problem.
Detailed explanation:
vectors $\vec{a}$, $\vec{b}$, satisfying $\vec{a}+\vec{b}=(2,3), \vec{a}-\vec{b}=(-2,1)$,

Therefore, $|\vec{a}|^2-|\vec{b}|^2=(\vec{a}+\vec{b})\cdot(\vec{a}-\vec{b})=2\times(-2)+3\times1=-1$.
Hence, the answer is B.

4. Among the following functions, which one is monotonically increasing on the interval $(0, +\infty)$?

A. $f(x) = -\ln x$

B. $f(x) = \frac{1}{2^x}$

C. $f(x) = -\frac{1}{x}$

D. $f(x) = 3^{|x-1|}$

4. Answer: C

Analysis: Utilize the monotonicity of basic elementary functions and combine the monotonicity of composite functions to judge ABC, and use counterexamples to exclude D.

Detailed explanation:

For A, because y = ln x is monotonically increasing on (0, +∞), y = -x is monotonically decreasing on (0, +∞).

Therefore, f(x) = -ln x is monotonically decreasing on (0, +∞), so A is incorrect.

For B, because y = 2^x is monotonically increasing on (0, +∞), $y = \frac{1}{2^x}$ is monotonically decreasing on (0, +∞). Therefore, f(x) = 1/2^x is monotonically decreasing on (0, +∞), so B is incorrect.

For C, because y = -1/x is monotonically decreasing on (0, +∞), y = -x is monotonically decreasing on (0, +∞). Therefore, f(x) = -1/x is monotonically increasing on (0, +∞), so C is correct.

For D, because f(1) = 3^|1-1| = 3^0 = 1, f(2) = 3^|2-1| = 3^1 = 3.

Obviously, f(x) = 3^|x-1| is not monotonic on (0, +∞), so D is incorrect.

5. Find the coefficient of x in the expansion of $\left(2x - \frac{1}{x}\right)^5$

- A. -80
- B. -40
- C. 40
- D. 80

5. Answer: D

Analysis: Write out the general term of the expansion of $\left(2x - \frac{1}{x}\right)^5$.

Detailed explanation:

The general term of the expansion of $\left(2x - \frac{1}{x}\right)^5$ is

$T_{r+1} = C_5^r (2x)^{5-r} \left(-\frac{1}{x}\right)^r = (-1)^r 2^{5-r} C_5^r x^{5-2r}$.

Let 5 - 2r = 1, we get r = 2.

Therefore, the coefficient of x in the expansion of $\left(2x - \frac{1}{x}\right)^5$ is $(-1)^2 2^{5-2} C_5^2 = 80$.

Hence, the answer is D.

Note: This question tests the application of the binomial theorem.

6. Given parabola C: $y^2 = 8x$ with focus F. Point M is on C. If the distance from M to the line x = -3 is 5, FInd |MF|

Analysis: Use the definition of a parabola to solve the problem.

Detailed explanation:

Because the parabola C: $y^2 = 8x$ has focus F(2, 0) and directrix x = -2, and point M is on C,

Therefore, the distance from M to the directrix x = -2 is |MF|.

Also, the distance from M to the line $x = -3$ is 5,
So $|MF| + 1 = 5$, hence $|MF| = 4$.

In △ABC, (a+c)(sinA-sinC) = b(sinA-sinB), find ∠C

- A. $\pi/6$
- B. $\pi/3$
- C. $2\pi/3$
- D. $5\pi/6$

Answer: B
Analysis: Utilize the sine law for side-angle transformation and the cosine law to solve the problem.
Detailed explanation:
Because $(a+c)(\sin A-\sin C) = b(\sin A-\sin B)$,
Therefore, by the sine law, $(a+c)(a-c) = b(a-b)$, which means $a^2-c^2=ab-b^2$,
Then $a^2+b^2-c^2=ab$, so $\cos C = (a^2+b^2-c^2)/(2ab) = 1/2$,
And $0<C<\pi$, so $C = \pi/3$.
Therefore, the answer is B.

8. If xy ≠ 0, then "x + y = 0" is a ( ) condition for "y/x + x/y = -2".

- A. Sufficient but not necessary condition
- B. Necessary but not sufficient condition
- C. Sufficient and necessary condition
- D. Neither sufficient nor necessary condition

8.【答案】C

【分析】解法一：由 $\frac{x}{y}+\frac{y}{x}=-2$ 化简得到 $x+y=0$ 即可判断；解法二：证明充分性可由 $x+y=0$ 得到 $x=-y$，代入 $\frac{x}{y}+\frac{y}{x}$ 化简即可，证明必要性可由 $\frac{x}{y}+\frac{y}{x}=-2$ 去分母，再用完全平方公式即可；解法三：证明充分性可由 $\frac{x}{y}+\frac{y}{x}$ 通分后用配凑法得到完全平方公式，再把 $x+y=0$ 代入即可，证明必要性可由 $\frac{x}{y}+\frac{y}{x}$ 通分后用配凑法得到完全平方公式，再把 $x+y=0$ 代入，解方程即可.

8. Answer: C
Analysis: Method 1: From the equation y/x + x/y = -2, we can directly obtain x + y = 0 to determine the condition. Method 2: To prove sufficiency, we can obtain x + y = 0 from y/x + x/y = -2 by using the method of completing the square.
Method 3: **Proof of sufficiency:** By finding a common denominator and using the method of completing the square, we can obtain a perfect square formula. Then, substituting x + y = 0, we can solve the equation.
Equation:
yx+xy=−2
Solution:

1. **Find a common denominator:**

Multiply the first fraction by x/x and the second fraction by y/y:
xyx2+xyy2=−2

2. **Combine the fractions:**

xyx2+y2=−2

3. **Cross-multiply:**

x2+y2=−2xy

4. **Rearrange the equation:**

$x2+2xy+y2=0$

5. **Factor the perfect square trinomial:**

$(x+y)2=0$

6. **Solve for x + y:**

$x+y=0$

Therefore, the solution to the equation is **x + y = 0**.

9. The sloped roof is one of the traditional architectural forms in China, containing rich mathematical elements. Installing light strips can outline the architectural contour and show the beauty of the shape. As shown in the figure, a certain sloped roof can be regarded as a five-sided solid, in which two sides are congruent isosceles trapezoids and two sides are congruent isosceles triangles. If AB = 25m, BC = AD = 10m, and the tangent of the angle between the plane of the isosceles trapezoid, the plane of the isosceles triangle, and the plane ABCD is $\frac{\sqrt{14}}{5}$**,**
Find the sum of all the edges of the five-sided solid

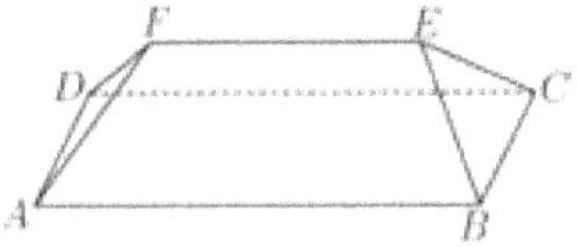

A. 102m B. 112m

C. 117m D. 125m

9. Answer: C

Analysis: First, according to the definition of the line-plane angle, we can obtain tan∠EMO = tan∠LEGO = √14/5. From this, we can find EO, EG, EB, and EF, and finally add up all the edge lengths to get the answer.

Detailed explanation: According to the question, the angles between the plane of the isosceles trapezoid, the plane of the isosceles triangle, and the base plane are ∠EMO and ∠EGO, respectively.

Based on the properties of the roof, the angles between the inclined plane and the base plane are ∠EMO and ∠EGO. Therefore, tan ∠EMO = tan ∠EGO = √14/5.
Because EO is perpendicular to plane ABCD and BC is on plane ABCD, so EO is perpendicular to BC.

Because EG is perpendicular to BC, and EO, EG are on plane, $EO \cap EG = E$,

So that BC is perpendicular plane EOG, because OG is on plane EOG, so $BC \perp OG$,

Similarly, OM is perpendicular to BM, and BM is perpendicular to BG, so quadrilateral OMBG is a rectangle.
Therefore, from BC = 10, we get OM = 5, so EO = √14, so OG = 5.

In right triangle EOG, $EG = \sqrt{EO^2 + OG^2} = \sqrt{\left(\sqrt{14}\right)^2 + 5^2} = \sqrt{39}$

In right triangle EBG, $EB = \sqrt{EG^2 + BG^2} = \sqrt{\left(\sqrt{39}\right)^2 + 5^2} = 8$,
Jm lso, because EF = AB - 5 - 5 = 25 - 5 - 5 = 15.
The sum of all edge lengths is 2 × 25 + 2 × 10 + 15 + 4 × 8 = 117m.
Therefore, the answer is C.

- 绝密启用前 (Top Secret: Activate Before Use)
- **2019 年普通高等学校招生全国统一考试** (2019 National Unified Entrance Examination for Ordinary Colleges and Universities)
- **理科数学** (Mathematics for Science Students)
- **注意事项:** (Notes)
  1. **答卷前,考生务必将自己的姓名,考生号等填写在答题卡和试卷指定位置上。** (Before answering the questions, candidates must fill in their name, candidate number, etc. on the designated positions of the answer sheet and the test paper.)
  2. **回答选择题时,选出每小题答案后,用铅笔把答题卡上对应题目的答案标号涂黑,如需改动,用橡皮擦干净后,再选涂其他答案标号。回答非选择题时,将答案写在答题卡上。写在本试卷上无效。** (When answering multiple-choice questions, after selecting the answer to each question, fill in the corresponding answer number on the answer sheet with a pencil. If you need to make changes, erase it cleanly and then fill in the other answer number. When answering non-multiple-choice questions, write the answers on the answer sheet. Writing on this test paper is invalid.)
  3. **考试结束后,将本试卷和答题卡一并交回。** (After the exam, return both this test paper and the answer sheet.)

Bottom Left Corner:

- **一、选择题:本题共 12 小题,每小题 5 分,共 60 分。在每小题给出的四个选项中,只有一项是符合题目要求的。** (Section 1: Multiple Choice Questions: There are 12 questions in this section, with 5 points for each question, totaling 60 points. Among the four options given for each question, only one is correct.)

Questions:

1. **已知集合 $M=\{x|-4<x<2\}$, $N=\{x|x^2-x-6<0\}$,则 $M\cap N$** (Given sets $M=\{x|-4<x<2\}$, $N=\{x|x^2-x-6<0\}$, then $M\cap N$)
   - A. $\{x|-4<x<3\}$
   - B. $\{x|-4<x<-2\}$
   - C. $\{x|-2<x<2\}$
   - D. $\{x|2<x<3\}$
2. **设复数 z 满足 2-i=1,在复平面内对应的点为(x,y),则** (Let complex number z satisfy 2-i=1, and the corresponding point in the complex plane is (x,y), then)
   - A. $(x+1)^2+y^2=1$
   - B. $(x-1)^2+y^2=1$
   - C. $x^2+(y-1)^2=1$
   - D. $x^2+(y+1)^2=1$
3. **已知 $a=\log_2 0.2$, $b=2^{0.2}$, $c=0.2^3$,则** (Given $a=\log_2 0.2$, $b=2^{0.2}$, $c=0.2^3$, then)
   - A. $a<b<c$
   - B. $a<c<b$
   - C. $c<a<b$
   - D. $b<c<a$

5. The approximate graph of the $f(x)=\frac{\sin x+x}{\cos x+x^2}$ 在$[-\pi, \pi]$ **in the interval [-π, π] is:**

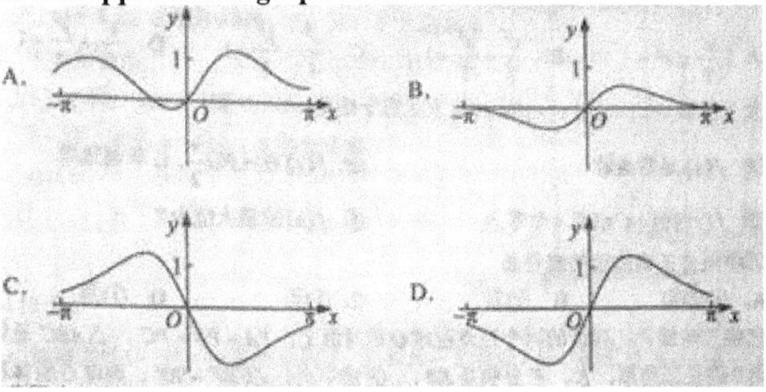

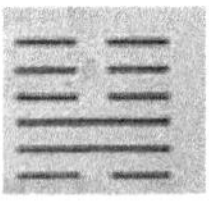

6.

In ancient China, the classic book "I Ching" used "gua" to describe the changes of all things. Each "gua" consists of 6 yao arranged from bottom to top. Yao is divided into yang yao "—" and yin yao " - - ". The right figure is a gua. If you randomly select a gua from all guas, the probability that the gua has exactly 3 yang yao is:

A. $\frac{5}{16}$ B. $\frac{11}{32}$ C. $\frac{21}{32}$ D. $\frac{11}{16}$

7. Given non-zero vectors a, b satisfying |a| = 2|b|, and (a-b)⊥b, then the angle between a and b is:

A. $\frac{\pi}{6}$ B. $\frac{\pi}{3}$ C. $\frac{2\pi}{3}$ D. $\frac{5\pi}{6}$

Question 10

Given an ellipse with foci F1(-1, 0) and F2(1, 0), and a straight line intersecting the ellipse at points A and B. If $|AF_2| = 2|F_2B|$, $|AB| = |BF_1|$, find the equation of the ellipse.

A. $\frac{x^2}{2}+y^2=1$ B. $\frac{x^2}{3}+\frac{y^2}{2}=1$ C. $\frac{x^2}{4}+\frac{y^2}{3}=1$ D. $\frac{x^2}{5}+\frac{y^2}{4}=1$

Question 11

Given the function f(x) = sin|x| + |sinx|, determine which of the following statements are true:

- f(x) is an even function.
- f(x) is increasing on the interval (0, π/2).
- f(x) has 4 zeros on the interval [-π, π].
- The maximum value of f(x) is 2.

A. ①②④ B. ②④ C. ①④ D. ①③

Question 12

Given a triangular pyramid P-ABC with all four vertices on the surface of a sphere. PA = PB = PC, triangle ABC is an equilateral triangle with side length 2. E and F are the midpoints of PA and AB respectively, and angle CEF = 90 degrees. Find the volume of the sphere.

Fill-in-the-blank Questions

A. $8\sqrt{6}\pi$ B. $4\sqrt{6}\pi$ C. $2\sqrt{6}\pi$ D. $\sqrt{6}\pi$

Question 13

The equation of the tangent line to the curve y = 3(x^2 + x)e^x at point (0, 0) is:

Question 14

Let Sn be the sum of the first n terms of the geometric sequence {an}. If a1 = 1/3, a2 = -1/9, and Sn = 5, find n.

Question 15

Team A and Team B are playing a best-of-seven basketball final. The schedule is AA-BB-AB-BA. The probability of Team A winning at home is 0.6, and the probability of winning away is 0.5. Assuming the results of each game are independent, find the probability of Team A winning 4-1.

Question 16

Given a hyperbola x^2/a^2 - y^2/b^2 = 1 (a > 0, b > 0) with left and right foci F1 and F2 respectively. A straight line passing through F1 intersects the asymptotes of the hyperbola at points A and B. If $\overrightarrow{F_1A}=\overrightarrow{AB}$, $\overrightarrow{F_1B}\cdot\overrightarrow{F_2B}=0$, find the eccentricity of the hyperbola.

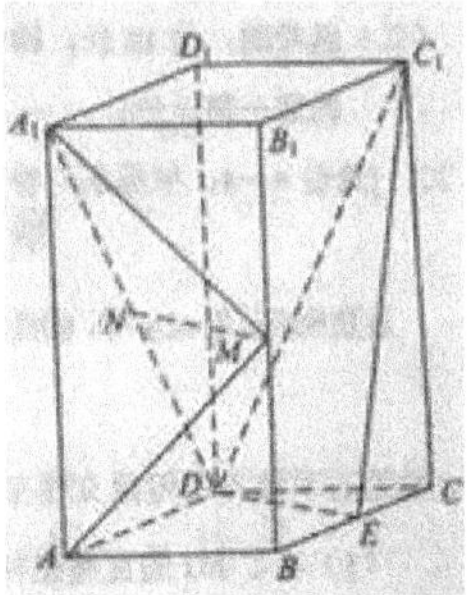

Question 18 (12 points)

As shown in the figure, $ABCD-A_1B_1C_1D_1$ is a right quadrilateral prism with a rhombus base. AB = 4, AD = 2, ∠BAD = 60°, E, M, and N are the midpoints of BC, BB_1 'and A_1D respectively.
(1) Prove: MN is parallel to plane CDE.
(2) Find the sine value of the dihedral angle $A-MA_1-N$.

Question 19 (12 points)
Given parabola C: $y^2 = 3x$ with focus F, a straight line with slope 3/2 intersects C at points A and B, and intersects the x-axis at point P.
(1) If |AF| + |BF| = 4, find the equation of the straight line.
(2) If $\overrightarrow{AP}=3\overrightarrow{PB}$, find |AB|.

Question 20 (12 points)
Given function f(x) = sinx - ln(1+x), f'(x) is the derivative of f(x). Prove:
(1) f'(x) has a unique maximum value point in the interval $(-1,\frac{\pi}{2})$.
(2) f(x) has exactly 2 zeros.

22. [Elective 4-4: Coordinate Systems and Parametric Equations] (10 points)

In the rectangular coordinate system xOy, the parametric equation of curve C is: $\begin{cases} x=\frac{1-t^2}{1+t^2}, \\ y=\frac{4t}{1+t^2} \end{cases}$ (t is a parameter). Taking the origin as the pole and the positive x-axis as the polar axis, establish a polar coordinate system. The polar equation of line l is $2\rho\cos\theta+\sqrt{3}\rho\sin\theta+11=0$.
(1) Find the rectangular coordinate equations of C and l.
(2) Find the minimum distance from points on C to line l.

23. [Elective 4-5: Inequality Selection] (10 points)
Given that a, b, and c are positive numbers and abc = 1. Prove:
(1) $\frac{1}{a}+\frac{1}{b}+\frac{1}{c}\leq a^2+b^2+c^2$;
(2) $(a+b)^3+(b+c)^3+(c+a)^3\geq 24$.

The answers to the quesiton above:

1. C	2. C	3. B	4. B	5. D	6. A
7. B	8. A	9. A	10. B	11. C	12. D

13. $y=3x$ 14. 121 [illegible] 15. 0.18 16. 2

Question 17. Solution:
(1) From the given information, we have $\sin^2B + \sin^2C - \sin A\sin B\sin C = 0$. Therefore, by the sine law, we get $b^2 + c^2 - a^2 = bc$.
From the cosine law, we get $\cos A = (b^2 + c^2 - a^2) / 2bc = 1/2$.
Because 0° < A < 180°, therefore A = 60°.
(2) From (1), we know B = 120° - C. From the given information and the sine law, we get $\sqrt{2}\sin A + \sin(120° - C) = 2\sin C$.

Then $\frac{\sqrt{6}}{2}+\frac{\sqrt{3}}{2}\cos C+\frac{1}{2}\sin C=2\sin C$, which gives cos(C + 60°) = -√2/2.
Since 0° < C < 120°, therefore sin(C + 60°) = √2/2. Thus,

$$\begin{aligned}\sin C &= \sin(C+60^\circ-60^\circ)\\ &= \sin(C+60^\circ)\cos 60^\circ-\cos(C+60^\circ)\sin 60^\circ\\ &= \frac{\sqrt{6}+\sqrt{2}}{4}.\end{aligned}$$

Question 18. Solution:

(1) Connect B_1C, ME. Because M and E are the midpoints of BB1 and BC respectively, so $ME \parallel B_1C$, and $ME=\frac{1}{2}B_1C$. Also, because N is the midpoint of A_1D, so $ND=\frac{1}{2}A_1D$.

From the given conditions, $A_1B_1 \underline{\parallel} DC$, so we get $B_1C \underline{\parallel} A_1D$, so $ME \underline{\parallel} ND$. Therefore, quadrilateral MNDE is a parallelogram, MN // ED, and MN is on plane EDC, so MN // plane C_1DE.

(2) From the known conditions, DE ⊥ DA. Taking point D as the origin, the direction of $\overrightarrow{DA}$ as the positive direction of the x-axis, establish a three-dimensional rectangular coordinate system D-xyz as shown in the figure. Then

$A(2,0,0)$, $A_1(2,0,4)$, $M(1,\sqrt{3},2)$, $N(1,0,2)$, $\overrightarrow{A_1A}=(0,0,-4)$, $\overrightarrow{A_1M}=(-1,\sqrt{3},-2)$, $\overrightarrow{A_1N}=(-1,0,-2)$, $\overrightarrow{MN}=(0,-\sqrt{3},0)$.

Let m = (x, y, z) be the normal vector of plane A, MA. Then...

$\begin{cases} m\cdot\overrightarrow{A_1M}=0,\\ m\cdot\overrightarrow{A_1A}=0.\end{cases}$ so $\begin{cases}-x+\sqrt{3}y-2z=0,\\ -4z=0,\end{cases}$ then $m=(\sqrt{3},1,0)$.

Let n = (p, q, r) be the normal vector of plane A, MN, then...

$\begin{cases} n\cdot\overrightarrow{MN}=0,\\ n\cdot\overrightarrow{A_1N}=0.\end{cases}$ so $\begin{cases}-\sqrt{3}q=0,\\ -p-2r=0.\end{cases}$ then $n=(2,0,-1)$. therefore

$\cos\langle m,n\rangle=\frac{m\cdot n}{|m||n|}=\frac{2\sqrt{3}}{2\times\sqrt{5}}=\frac{\sqrt{15}}{5}$. **Therefore, the sine of the dihedral angle $A-MA_1-N$ is √10/5.**

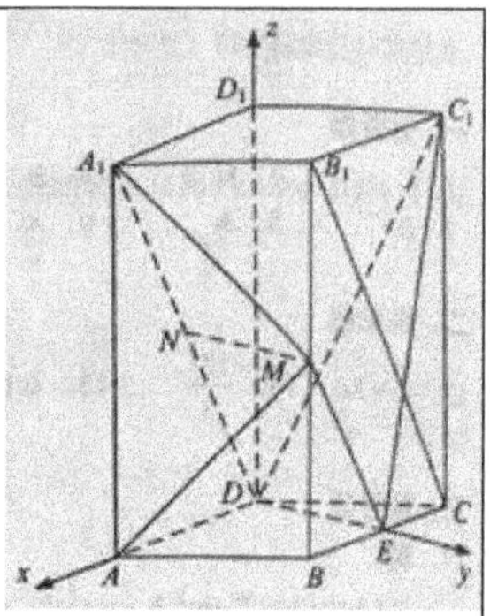

19 solution:

Let straight linel: $y=\frac{3}{2}x+t$, $A(x_1,y_1)$, $B(x_2,y_2)$. From the condition in the problem we know: $F(\frac{3}{4},0)$, so $|AF|+|BF|=x_1+x_2+\frac{3}{2}$, from the condition in the problem:, we know: $x_1+x_2=\frac{5}{2}$.

From $\begin{cases}y=\frac{3}{2}x+t,\\ y^2=3x\end{cases}$ we know: $9x^2+12(t-1)x+4t^2=0$, then $x_1+x_2=-\frac{12(t-1)}{9}$, so that: $-\frac{12(t-1)}{9}=\frac{5}{2}$, we get: $t=-\frac{7}{8}$, theerefore, the function of l: $y=\frac{3}{2}x-\frac{7}{8}$.

20 solution:

Let $g(x)=f'(x)$, then $g(x)=\cos x-\frac{1}{1+x}$, $g'(x)=-\sin x+\frac{1}{(1+x)^2}$. when $x\in(-1,\frac{\pi}{2})$

g'(x) is monotonically decreasing.and $g'(0)>0$, $g'(\frac{\pi}{2})<0$, we get $g'(x)$ 在 $(-1,\frac{\pi}{2})$ has only one zero α.

then when $x\in(-1,\alpha)$ $g'(x)>0$; when $x\in(\alpha,\frac{\pi}{2})$ $g'(x)<0$. therefore, g(x) is monotonically increasing at $(\alpha,\frac{\pi}{2})$, so g(x) has a maximum point at $(-1,\frac{\pi}{2})$, which means $f'(x)$ has maximum point in $(-1,\frac{\pi}{2})$

2. The domain of f(x) is (-1, +∞).
(i) When x ∈ (-1, 0], by (1), f'(x) is monotonically increasing on (-1, 0), and f'(0) = 0. Therefore, when x ∈ (-1, 0), f'(x) < 0, so f(x) is monotonically decreasing on (-1, 0). Also, f(0) = 0, so x = 0 is the only zero point of f(x) on (-1, 0].
(ii) When x ∈ (0, π/2), by (1), f'(x) is monotonically increasing on (0, a) and monotonically decreasing on (a, π/2).
And f'(0) = 0, f'(π/2) < 0, so there exists β ∈ (a, π/2) such that f'(β) = 0. And when x ∈ (0, β), f'(x) > 0; when x ∈ (β, π/2), f'(x) < 0. So f(x) is monotonically increasing on (0, β) and monotonically decreasing on (β, π/2).
Also, f(0) = 0, f(π/2) = 1 - ln(1 + π/2) > 0. So when x ∈ (0, π/2], f(x) > 0. Therefore, f(x) has no zero points on (0, π/2].

(iii) When x is in the interval (π/2, e), f'(x) is less than 0. Therefore, f(x) is monotonically decreasing on (π/2, π). And f(π/2) is greater than 0.
f(x) is less than 0. Therefore, f(x) has exactly one zero point in the interval (π/2, e).
(iv) When x is in the interval (π, +∞), ln(x+1) is greater than 1. Therefore, f(x) is less than 0. Thus, f(x) has no zero points in the interval (π, +∞).
In summary, f(x) has exactly two zero points.

21.
(1) All possible values of X are -1, 0, 1.
$P(X=-1)=(1-\alpha)\beta$,
$P(X=0)=\alpha\beta+(1-\alpha)(1-\beta)$,
$P(X=1)=\alpha(1-\beta)$.
Therefore, the distribution of X is...

X	-1	0	1
P	$(1-\alpha)\beta$	$\alpha\beta+(1-\alpha)(1-\beta)$	$\alpha(1-\beta)$

(2) By(1) we know: $a=0.4$, $b=0.5$, $c=0.1$.
Therefore, $p_i=0.4p_{i-1}+0.5p_i+0.1p_{i+1}$, so $0.1(p_{i+1}-p_i)=0.4(p_i-p_{i-1})$, which means: $p_{i+1}-p_i=4(p_i-p_{i-1})$. and as $p_1-p_0=p_1\neq 0$, so $\{p_{i+1}-p_i\}(i=0,1,2,\cdots,7)$ is a geometric sequence with a common ratio of 4 and the first term of p.

(ii) by (i) we know:
$$\begin{aligned} p_8 &= p_8-p_7+p_7-p_6+\cdots+p_1-p_0+p_0 \\ &= (p_8-p_7)+(p_7-p_6)+\cdots+(p_1-p_0) \\ &= \frac{4^8-1}{3}p_1. \end{aligned}$$
as $p_8=1$, so $p_1=\frac{3}{4^8-1}$, therefore:
$$\begin{aligned} p_4 &= (p_4-p_3)+(p_3-p_2)+(p_2-p_1)+(p_1-p_0) \\ &= \frac{4^4-1}{3}p_1 \\ &= \frac{1}{257}. \end{aligned}$$
P represents the probability of ultimately believing that drug A is more effective. According to the calculation results, when the cure rate of drug A is 0.5 and the cure rate of drug B is 0.8, the probability of believing that drug A is more effective is p = 1/257 ≈ 0.0039. At this time, the probability of drawing an incorrect conclusion is very small, indicating that this experimental design is reasonable.

22, solution:

As $-1<\frac{1-t^2}{1+t^2}\leq 1$，且 $x^2+(\frac{y}{2})^2=(\frac{1-t^2}{1+t^2})^2+\frac{4t^2}{(1+t^2)^2}=1$， so the Cartesian Coordinate Equation of C is $x^2+\frac{y^2}{4}=1\ (x\neq -1)$. and the Cartesian Coordinate Equation of l is $2x+\sqrt{3}y+11=0$.

(3) By (1) we can le the Parametric equations of C to be $\begin{cases} x=\cos\alpha, \\ y=2\sin\alpha \end{cases}$ (α is the parametric, $-\pi<\alpha<\pi$).

The distance of points on C to l: $\frac{|2\cos\alpha+2\sqrt{3}\sin\alpha+11|}{\sqrt{7}}=\frac{4\cos(\alpha-\frac{\pi}{3})+11}{\sqrt{7}}$. when $\alpha=-\frac{2\pi}{3}$ the minimum value of $4\cos(\alpha-\frac{\pi}{3})+11$ is 7, so the distance of points on C to l is $\sqrt{7}$.

23 solution:

(1) Because $a^2+b^2\geq 2ab$， $b^2+c^2\geq 2bc$， $c^2+a^2\geq 2ac$, and abc=1, so we have: $a^2+b^2+c^2\geq ab+bc+ca=\frac{ab+bc+ca}{abc}=\frac{1}{a}+\frac{1}{b}+\frac{1}{c}$. so $\frac{1}{a}+\frac{1}{b}+\frac{1}{c}\leq a^2+b^2+c^2$.

(2) because a， b， c are all positive integers, and abc=1 so we have:

$$\begin{aligned}(a+b)^3+(b+c)^3+(c+a)^3 &\geq 3\sqrt[3]{(a+b)^3(b+c)^3(a+c)^3} \\ &=3(a+b)(b+c)(a+c) \\ &\geq 3\times(2\sqrt{ab})\times(2\sqrt{bc})\times(2\sqrt{ac}) \\ &=24.\end{aligned}$$

so that $(a+b)^3+(b+c)^3+(c+a)^3\geq 24$.

Book Review: ***China College Entrance Examination Mathematics Questions***

Part of the "How Chinese Students Achieve High Score in Mathematics" Series

Collected & Translated by Haiqing Hua

The "China College Entrance Examination Mathematics Questions" book is a remarkable addition to the *How Chinese Students Achieve High Score in Mathematics* series, meticulously compiled and translated by Haiqing Hua. This book offers an authentic glimpse into the rigorous mathematical challenges Chinese students face during their college entrance examinations, known as the Gaokao.

Content Overview:

The book features a collection of mathematics questions from several significant years and variations of the Gaokao, including:

- **2022 National Unified Entrance Examination**
- **2024 National Volume I, Mathematics**
- **2024 National College Entrance Examination (New Curriculum I)**
- **2024 National Unified Entrance Examination for Ordinary Higher Education**
- **2023 Beijing College Entrance Exam**
- **2019 National Unified Entrance Examination for Ordinary Colleges and Universities**

Each set of questions is presented in its original form and translated into English, making it accessible to an international audience eager to understand the depth and complexity of Chinese mathematics education. By providing these examples, the book demonstrates the high level of analytical thinking, problem-solving, and mathematical understanding that is expected of Chinese students.

Strengths:

1. **Authenticity and Relevance:** The book is a valuable resource for educators, students, and researchers interested in global education standards. By translating real examination questions, Haiqing Hua ensures that readers get a true representation of the challenges Chinese students face.
2. **Insight into Chinese Education:** The Gaokao is one of the most challenging college entrance exams in the world, and this book sheds light on why Chinese students consistently excel in mathematics. The collection serves as a testament to the effectiveness of China's education system in cultivating high mathematical competence.
3. **Bilingual Accessibility:** The dual-language format not only helps non-Chinese speakers gain insight into the Gaokao but also serves as a learning tool for students who wish to practice their mathematical skills in both English and Chinese.

Potential Areas for Improvement:

1. **Explanations and Solutions:** While the book excels in providing a collection of questions, it could be enhanced by offering detailed solutions or explanations for each problem. This would be particularly helpful for students and educators who are not only interested in the questions but also in the methodologies behind solving them.
2. **Contextual Background:** Including a brief overview of the structure and significance of the Gaokao, as well as how mathematics fits into the overall examination, could provide readers with more context and appreciation for the material presented.

Conclusion:

China College Entrance Examination Mathematics Questions is an excellent resource for anyone interested in understanding how Chinese students achieve high scores in mathematics. Through a carefully curated selection of questions from recent years, Haiqing Hua opens a window into the high standards of mathematical education in China. While the book is primarily a collection of problems, its potential as a learning and teaching tool is immense. With the addition of solutions and more contextual information, it could become an even more valuable resource for educators and students worldwide.

www.ingramcontent.com/pod-product-compliance
Lightning Source LLC
Chambersburg PA
CBHW050325230526
45471CB00005B/2350
9798335950060